SpringerBriefs in Environmental Science

SpringerBriefs in Environmental Science present concise summaries of cutting-edge research and practical applications across a wide spectrum of environmental fields, with fast turnaround time to publication. Featuring compact volumes of 50 to 125 pages, the series covers a range of content from professional to academic. Monographs of new material are considered for the SpringerBriefs in Environmental Science series.

Typical topics might include: a timely report of state-of-the-art analytical techniques, a bridge between new research results, as published in journal articles and a contextual literature review, a snapshot of a hot or emerging topic, an in-depth case study or technical example, a presentation of core concepts that students must understand in order to make independent contributions, best practices or protocols to be followed, a series of short case studies/debates highlighting a specific angle.

SpringerBriefs in Environmental Science allow authors to present their ideas and readers to absorb them with minimal time investment. Both solicited and unsolicited manuscripts are considered for publication.

More information about this series at http://www.springer.com/series/8868

Parmod Kumar · Surender Kumar
Laxmi Joshi

Socioeconomic and Environmental Implications of Agricultural Residue Burning

A Case Study of Punjab, India

Parmod Kumar
ADRTC, Institute for Social
  and Economic Change
Bengaluru
Karnataka
India

Surender Kumar
Department of Business Economics
University of Delhi
New Delhi
India

Laxmi Joshi
Department of Agriculture and Rural
  Development
National Council of Applied Economic
  Research
Parisila Bhawan
New Delhi
India

ISSN 2191-5547   ISSN 2191-5555   (electronic)
ISBN 978-81-322-2146-3   ISBN 978-81-322-2014-5   (eBook)
DOI 10.1007/978-81-322-2014-5

Library of Congress Control Number: 2014953254

Springer New Delhi Heidelberg New York Dordrecht London

Contents

**1 Problem of Residue Management Due to Rice Wheat
   Crop Rotation in Punjab**   1
   1.1 Agricultural Growth in Punjab...........................   1
   1.2 Agricultural Residue Burning and Its Management..............   3
   1.3 Main Objectives of the Study............................   10
   1.4 An Overview   10
   References..   12

2 The Extent and Management of Crop Stubble.   13
   2.1 Introduction   13
   2.2 The Produce of Crop Stubble and Its Burning.................   14
      2.2.1 Straw/Residue to Grain Ratio.......................   17
      2.2.2 Chemical Composition of Rice and Wheat Stubble........   18
   2.3 Volume of Pollution Caused by Crop Stubble Burning...........   19
   2.4 Effects of Crop Stubble Burning on the Fertility of the Soil   24
      2.4.1 International Experience...........................   25
   2.5 Health Impacts of Pollution Due to Residue Burning   25
   2.6 Management of Crop Stubble............................   26
      2.6.1 In Situ Incorporation   26
      2.6.2 Alternative Uses of Crop Stubble....................   27
      2.6.3 Cost of Alternate Uses   29
      2.6.4 End Use of Rice Residue in Different Districts of Punjab...   29
   2.7 Summary of the Chapter...............................   31
   References..   32

3 Valuation of the Health Effects   35
   3.1 Introduction   35
   3.2 Ambient Air Quality Level in Study Area   37
   3.3 Household Survey Design and Data.........................   38
   3.4 The Survey Results...................................   40

3.4.1   The Household and Farming Characteristics   40
        3.4.2   Management of Stubble Among the Selected Farmers   43
        3.4.3   The Effect of Crop Stubble Burning on Human Health.   48
    3.5   Methodology   56
        3.5.1   Theoretical Model   56
        3.5.2   Estimation Strategy   58
    3.6   The Model Results   60
        3.6.1   Welfare Loss.   62
        3.6.2   Increase in Medical Expenditure   62
        3.6.3   Opportunity Cost of Increase in Workdays Lost   63
    3.7   Summary of the Chapter.   64
    References.   66

4   Alternative Uses of Crop Stubble.   69
    4.1   Introduction   69
    4.2   Disposal Pattern of Paddy Straw   70
    4.3   Management of Agricultural Waste for Alternate Uses   71
        4.3.1   Use of Rice Residue as Fodder for Animals.   71
        4.3.2   Use of Crop Residue in Bio Thermal Power Plants   75
        4.3.3   Use of Rice Residue as Bedding Material for Cattle   77
        4.3.4   Use of Crop Residue for Mushroom Cultivation   77
        4.3.5   Use of Rice Residue in Paper Production   77
        4.3.6   Use of Rice Residue for Making Bio Gas   78
        4.3.7   In Situ.   78
        4.3.8   Incorporation of Paddy Straw in Soil.   78
        4.3.9   Production of Bio-oil from Straw and Other
                 Agricultural Wastes   79
    4.4   Agricultural Residues for Power Generation   79
        4.4.1   Energy Technologies   81
        4.4.2   Thermal Combustion   81
    4.5   Summary of the Chapter.   83
    References.   89

5   Environmental Legislations: India and Punjab   91
    5.1   Introduction   91
        5.1.1   Ministry of Environment and Forest   92
        5.1.2   Clean Technology Division   92
        5.1.3   Control of Pollution Division.   93
    5.2   Various Laws to Control Pollution in India   94
        5.2.1   Water Act (Prevention and Control of Pollution Act, 1974). . .   94
        5.2.2   Air Prevention and Control of Pollution Act, 1981   97
        5.2.3   The Environment Protection Act, 1986   100
        5.2.4   The Environment (Protection) Rules, 1986   102
        5.2.5   The National Environment Tribunal Act, 1995   103
        5.2.6   The National Environment Appellate Authority Act, 1997   104

	5.2.7	The Noise Pollution (Regulation and Control) Rules, 2000...	105
	5.2.8	Biological Diversity Act, 2002	106
5.3	Central Pollution Control Board (CPCB)		108
	5.3.1	Functions of the Central Board	108
	5.3.2	National Ambient Air Monitoring Programme (NAMP)...	109
	5.3.3	Water Quality Monitoring and Surveillance Programme...	109
5.4	Punjab Pollution Control Board (PPCB)		110
5.5	Punjab State Council for Science and Technology		112
5.6	Environment Division		113
5.7	Punjab Energy Development Agency (PEDA)		114
5.8	Punjab Biodiversity Board		115
5.9	Summary of the Chapter		115

6 Policies for Restricting the Agriculture Residue Burning in Punjab ... 117
6.1	Monitoring and Recording the Levels of Pollution in Punjab		118
6.2	Existing Policies to Control Air Pollution		122
	6.2.1	Punjab Pollution Control Board (PPCB)	122
	6.2.2	Agriculture Councils	123
	6.2.3	Punjab State Council for Science and Technology	124
	6.2.4	Department of Agriculture	124
	6.2.5	Punjab Energy Development Agency (PEDA)	124
	6.2.6	Department of Animal Husbandry	125
	6.2.7	Punjab Agricultural University	125
	6.2.8	Punjab State Farmers' Commission	126
	6.2.9	Department of Rural Development and Panchayats	126
	6.2.10	Agriculture Diversification	126
	6.2.11	Promotion of Zero Tillage	127
	6.2.12	Management of Agricultural Waste	127
	6.2.13	Utilization of Straw and Husk	127
	6.2.14	Use of Rice Residue as Fodder for Animals	127
	6.2.15	Use of Crop Residue in Bio Thermal Power Plants	128
	6.2.16	Use of Rice Residue as Bedding Material for Cattle	128
	6.2.17	Use of Crop Residue for Mushroom Cultivation	128
	6.2.18	Use of Rice Residue in Paper Production	129
	6.2.19	Use of Rice Residue for Making Bio Gas	129
	6.2.20	Other Measures	129
6.3	Summary of the Chapter		130
Reference			131

7 Concluding Remarks and Policy Recommendations ... 133
7.1	Introduction		133
7.2	Summary of the Findings		134
7.3	Policy Recommendations and Research Needs		138

Annexure ... 141

About the Authors

Parmod Kumar is Professor and Head (Director), Agricultural Development and Rural Transformation Centre, Institute for Social and Economic Change, Bengaluru, India. He has previously worked at the National Council of Applied Economic Research, New Delhi and the Institute of Economic Growth, Delhi. He obtained his postdoctorate as Sir Ratan Tata Fellow from the Institute of Economic Growth and doctorate from Jawaharlal Nehru University, New Delhi. He was fellow under the International Visitors Program sponsored by the US government. Professor Kumar has authored six research volumes and published more than 40 research articles in refereed national and international journals. He is leading several research projects sponsored by the Government of India and various international organizations. He is the managing editor of the *Journal of Social and Economic Change* and is on the editorial board of *Agricultural Situation in India* and the *Indian Journal of Agricultural Marketing*. He is a member of various committees of the Union and State governments. Professor Kumar was conferred the IDRC India Social Science Research Award for his work on the public distribution system.

Surender Kumar is Professor at the Department of Business Economics, University of Delhi, New Delhi, and is one of the lead authors for IPCC AR5. He has been a Visiting Fellow at the University of Illinois at Urbana-Champaign (USA) and Senior JSPS Fellow at the Yokohama National University Yokohama (Japan). Professor Kumar has authored four books: *Environmental and Economic Accounting for Industry* (Oxford University Press, New Delhi); *Economics of Sustainable Development: The Case of India* (Springer, New York); *Energy Prices and Induced Innovations* (VDM Verlag); and *Economics of Air Pollution* (VDM Verlag); and about 50 research papers in journals such as the *European Journal of Law and Economics*, *Ecological Economics*, *Economic Modelling*, *Environmental and Resource Economics*, *Environment and Development Economics*, *Resource and Energy Economics*, etc. He teaches courses in environmental economics and advanced econometrics.

Laxmi Joshi works at the Department of Agriculture and Rural Development of the National Council of Applied Economic Research, India. She has previously worked at the National Centre for Agricultural Economics and Policy Research, New Delhi as Senior Research Associate, and in the National Commission on Farmers, New Delhi as Research Officer. She has published a dozen papers in different subjects, such as agriculture policy for farmers, diversification, watershed development, and so on. Her research has been published in the *Indian Journal of Agricultural Economics*, *Economic and Political Weekly*, *International Water Organisation Colombo* (research report on water), etc. She has also co-edited a volume on livestock and different farming systems.

About the Book

This book discusses the important issue of the socioeconomic and environmental impacts of agricultural residue burning, common in agricultural practices in many parts of the world. In particular, it focuses on the pollution caused by rice residue burning using primary survey data from Punjab, India. It discusses emerging solutions to agricultural waste burning that are cost-effective in terms of both money and time. The burning of agricultural residue causes severe pollution in land, water, and air and contributes to increased ozone levels and climate change in the long term. However, appropriate assessments have not been undertaken so far to demonstrate the relevant impact of agriculture-based pollution, especially residue burning. This book addresses this gap in the literature. Punjab has been used as a case study as it is the chief granary of India, contributing to 27.2 % of the Indian national produce of rice and 43.8 % of wheat. It is presumed that the findings from this state will be useful not only for other agricultural areas in India, but across the world. This book, therefore, sensitizes policy makers, researchers, and students about the impacts of air pollution caused by agricultural residue burning—a subject not much dealt with in the literature—and provides a way forward.

Figures

Fig. 1.1   Productivity of major crops in Punjab 3
Fig. 3.1   Number of patients treated in the village
           dispensary Ajnauda Kalan............................... 53
Fig. 6.1   SPM/RSPM levels at different residential cum commercial
           locations in Punjab 119
Fig. 6.2 SO_2 levels at different residential cum commercial areas
           in Punjab. Status of Environment and Related Issues............ 120
Fig. 6.3 NO_2 levels at different residential cum commercial
           areas in Punjab 120

Tables

Table 1.1   Contribution of wheat and rice to the central pool by Punjab ...   2

Table 1.2   Cropping pattern in Punjab (GCA in thousand hectares)
(Area under various crops as percentage of gross
cropped area)......................................   5

Table 2.1   Total quantity of crop stubble generated in India as per
different studies....................................   15

Table 2.2   End use of stubble by the farmers......................   15

Table 2.3   Various studies reporting rice residue burnt in open
fields in Punjab....................................   16

Table 2.4   Residue to product ratio according to various studies.........   17

Table 2.5   Nutrient content of paddy straw and amounts
removed with one tonne of straw residue...................   18

Table 2.6   Nutrient content in rice residue........................   18

Table 2.7   Moisture and other factors in rice residue.................   18

Table 2.8   Chemical composition in rice and wheat straw.............   19

Table 2.9   Major pollutants emitted during crop residue burning.........   20

Table 2.10 National estimates of biomass burned and emission
of aerosols and trace gases for crop waste open burning.......   23

Table 2.11 Emission of trace gases from burning of rice and wheat residue...   23

Table 2.12 Total emission by burning of rice and wheat crop............   23

Table 2.13 Nutrient losses due to burning of rice residues
in Punjab, 2001–2002...............................   25

Table 2.14 Effect of crop residue management on organic C and total
N content of soil under the rice-wheat cropping system........   28

Table 2.15 Impact of different residue management practices
in Ludhiana (Punjab)...............................   28

Table 2.16 The effect of different crop residue management practices
on the soil..   28

Table 3.1   Descriptive statistics of emissions and metrological data.   38
Table 3.2   Total sown area and area under rice and wheat in the selected
             villages in the year 2001   39
Table 3.3   Household characteristics (%)   41
Table 3.4   Farm holding characteristics.   42
Table 3.5   Value of assets holding among the farmers (Rs per household). . .   43
Table 3.6   Input-output table (Rs per acre)   44
Table 3.7   The amount of stubble generated on the field
             and its alternate uses   45
Table 3.8   Residue removal practices in the field (% of households)   46
Table 3.9   Motivation for burning of crop residue (% of households).   46
Table 3.10   Average no of days available for the next crop when crop
             residue is removed by different practices   46
Table 3.11   The easiest and quickest way to get rid of the crop stubble
             (% of households).   47
Table 3.12   Households' perception about which method of crop stubble
             management gives them the maximum crop yield
             (% of households).   47
Table 3.13   If crop subtle incorporated in the soil, method used
             for incorporation (% of households).   48
Table 3.14   Additional fertilize use when crop stubble burning   48
Table 3.15   The effect of end use of straw on the amount
             of irrigation used per acre.   49
Table 3.16   The effect of end use of straw on the amount of fertilizer
             used per acre   51
Table 3.17   Percentage of household experiencing any problem due
             to smoke caused by crop stubble burning (mainly rice)   52
Table 3.18   Percentage of HH members suffering from the disease due
             to stubble burning   53
Table 3.19   Expenditure incurred due to problems faced during the crop
             stubble burning   54
Table 3.20   Medical expenses incurred due to health problem caused
             by crop stubble burning.   55
Table 3.21   Percentage of households saying yes to the following questions. . .   56
Table 3.22   Variables used in the analysis   60
Table 3.23   Tobit equation of total medical expenditure (left censured at 0). . .   61
Table 3.24   Poisson equation of workdays lost   61
Table 3.25   Welfare loss due to increased air pollution in rural Punjab   63
Table 4.1   End use of paddy straw.   70
Table 4.2   State-wise consumption of paddy (residue) per animal   72
Table 4.3   Status of different states about availability and requirement
             of fodder   73

Table 4.4   State-wise percentage of short fall of crop residue and greens. . .   74
Table 6.1   The rating scale for air quality index   118
Table 6.2   The descriptive categories for different exceedence
             indicator values   119
Table 6.3   National ambient air quality standards set by the CPCB   121

Annexure Tables

Table 3.26 Cropping pattern of selected farmers
            (percentage of gross cropped area) 65
Table 3.27 Are there any buyers of rice/wheat residue. 66

Chapter 1
Problem of Residue Management Due to Rice Wheat Crop Rotation in Punjab

Abstract Punjab agriculture supported by input and output price structure and superior yields of rice and wheat compared to other crops has virtually become a rice-wheat monoculture. The rice-wheat cycle has led to over exploitation of ground water resources in the state. Use of combined harvester has further exacerbated the problem of crop residue management as it leaves behind a large amount of rice residue to be burnt in the open fields. This study brings the problem of agriculture waste burning in the forefront. It tries to enumerate the amount of pollution being caused by rice residue burning and its adverse impact on human health.

Keywords Rice-wheat crop rotation · Combined harvester · Residue burning · Residue management

1.1 Agricultural Growth in Punjab

The Indian state of Punjab is known as the country's chief granary contributing almost one-fourth share of rice and more than one-third of wheat to the central pool. On an average, the state's share in the total production of wheat and rice in all-India stands about 20 and 10 %, respectively. Table 1.1 provides details of wheat and rice contribution by Punjab to the central pool over the last three decades. Further, above 95 % of food grains produced in the state go out of the state to feed food deficit areas through the public distribution system. The state agriculture is characterized as the backbone of the public distribution system and a strong base for the food security of the country.

Punjab made a commendable progress in the production of food grains in the post-green revolution period. Food grain production underwent a big jump from 3.16 million tonnes in 1960–1961 to 28.35 million tonnes in 2011–2012. The green revolution also known as the new agricultural strategy was marked with the arrival of new high yielding varieties of wheat, rice, maize and *bajra* (millet) and package of other inputs like chemical fertilizers, insecticides, pesticides and

© The Author(s) 2015

P. Kumar et al., *Socioeconomic and Environmental Implications of Agricultural Residue Burning*, SpringerBriefs in Environmental Science, DOI 10.1007/978-81-322-2014-5_1

Table 1.1 Contribution of wheat and rice to the central pool by Punjab

Period	Percent share of rice	Percent share of wheat
1980–1981	45.3	73.0
1990–1991	41.0	61.0
2000–2001	33.3	57.6
2005–2006	32.0	60.9
2006–2007	31.2	75.3
2007–2008	27.8	60.9
2008–2009	25.1	43.8
2009–2010	29.0	42.3
2010–2011	25.2	45.3
2011–2012	22.1	38.7
2012–2013	–	33.6

Source Statistical abstracts of Punjab, various years

assured irrigation facilities. Focusing on popularizing modern inputs and practices in the productive areas where the likelihood was more for the high yielding seeds to show results was the most important feature of this new strategy. Punjab with requisite irrigation and infrastructure facilities became a major beneficiary of this national strategy and has been shown as a showpiece of India's successful green revolution strategy. Short duration and high yielding varieties (HYV) of rice and wheat were introduced in Punjab to boost up the production of food grains. During the first decade of the green revolution, the technology was confined only to the wheat crop. A remarkable growth rate of 5 % was achieved by the state's agricultural sector since the beginning of the green revolution in the mid-1960s.

The decade from the mid-1970s to mid-1980s was characterized by the extension of new seed fertilizer technology from wheat to rice crop. Due to the input and output price structure and superior yields of rice and wheat as compared to other crops, Punjab agriculture has virtually become a rice-wheat monoculture. During 1966–1967, total area under rice was 0.29 million hectares, which increased to 2.82 million hectares by 2011–2012. There was also a substantial increase in the average rice productivity, which increased from 1,186 kg/ha in 1966–1967 to 3,741 kg/ha by 2011–2012 (Fig. 1.1). During the same period, the area under wheat crop increased from 1.61 million hectares to 3.51 million hectares and productivity from 1,544 to 4,898 kg/ha. Punjab has achieved a crop intensity of 188 % as against 138 in the country as a whole. The present level of consumption of fertilizer (NPK) is 244 kg/ha as compared to the Indian average of 144 kg/ha. Similarly, Punjab has 98 % of high yielding variety coverage, which is the highest among the Indian states. About 18 % of the total tractors in India are in Punjab. Production is supported by about 97 % irrigation coverage with 970,139 tube wells.

Given tremendous achievements in the past, however serious concerns are now emerging about the future prospects of Punjab's agricultural sector. Agricultural growth slowed substantially in the 1990s. Agricultural output grew at a trend rate

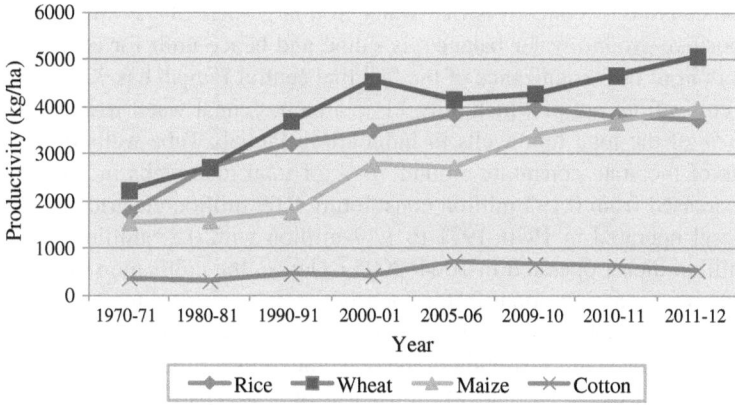

Fig. 1.1 Productivity of major crops in Punjab

of 2.6 % per annum in the 1990s compared to the all India average of 3.2 % and relative to a growth rate of 5 % per annum in Punjab in the 1980s. Productivity of rice appears to be reaching to plateau. The main rice-wheat tract of central Punjab also experienced a decrease in growth with total factor productivity (TFP) growth coming down to 0.07 % per annum mainly due to negative TFP growth in rice (Singh and Hussain 2002).

1.2 Agricultural Residue Burning and Its Management

The growth story of Punjab agriculture was accompanied by its negative environmental concerns. One of the concerns is about the over exploitation of ground water resources of the state. Punjab now has the highest percentage of ground water exploitation in the country and also the largest percentage of over exploited and dark blocks. As per the guidelines of Ground Water Resources Estimation Committee (GEC), the present ground water development (ratio of gross ground water draft for all uses to net ground water availability) in the state is 145 % as on March 2004. As per latest data provided by Central Ground Water Board, (Government of India 2011) and Department of Irrigation Punjab,[1] out of 137 blocks of the state, 103 blocks are overexploited, 5 blocks are critical, 4 blocks are semi critical and only 25 blocks are in safe category. All the blocks of various districts like Amritsar (16 blocks), Jalandhar (10 blocks), Moga (5 blocks), Kapurthala (5 blocks), Sangrur (12 blocks), Fatehgarh Sahib (5 blocks), Patiala (8 out of 9 blocks) and Ludhiana (9 out of 10 blocks) have been found to be overexploited, leading to sharp depletion of the water table in these districts.

---

[1] Report on dynamic ground water resources of Punjab, 2005, Government of Punjab.

Another issue of concern is that water in a large part of the area, which indicates positive ground water balance, is saline and hence unfit for consumption. It is important to take cognizance of the fact that central Punjab has 72 % area under paddy cultivation, out of which only 21 % area has canal water irrigation facility. Over 6 % of the total tube wells in India are in Punjab. Tube wells in the central districts of the state constitute around 70 % of total tube wells in Punjab, which have increased from 0.192 million constituting 0.09 million electric and 0.10 million diesel operated in 1970–1971 to 1.17 million with 0.88 million electric and 0.29 million diesel operated in 2004–2005.[2] Hence, the cultivation of high water-demand crops is an important factor contributing towards declining water levels in Punjab. It can be observed that the present grim scenario of ground water in different regions of the state is essentially the outcome of unscrupulous production practices leading to excessive and irrational use of water particularly for rice crop. Other factors include less than required availability of surface water, free power supply to the agricultural sector, support prices and procurement facilities for only some crops and disproportionate installation of tube wells by farmers.

Further, in the past two to three decades, intensive agricultural practices have put a tremendous pressure on the soils and resulted in steady decline in its fertility and nutrient availability both with respect to macro and micronutrients. Both, rice and wheat have high nutritional requirements and the double cropping of this system has been heavily depleting the nutrient contents of soil. For example, a rice-wheat sequence that yields 7 t/ha of rice and 5 t/ha of wheat removes more than 300 kg nitrogen, N, 30 kg phosphorus, P and 300 kg of potassium, K per hectare from the soil. Even with the recommended rate of fertilization in this cropping pattern, a negative balance of primary nutrients still exists (Benbi et al. 2006). Moreover, the partial factor productivity of NPK in Punjab has also dropped from 80.9 in 1966–1967 to 16.0 in 2003–2004. Hence, farmers in the state have been applying higher and higher doses of major nutrients, especially nitrogen for sustaining adequate production levels. Extensive use of nitrogenous fertilizers and pesticides has also led to increasing nitrate concentration and accumulation of pesticide residues in soil, water, food, feed and other agricultural produce often above tolerance limits.

Following the success of the high yielding varieties, there was introduction of rice-wheat cropping pattern in Punjab. It covers more than 2.6 million hectares or 60 % of the total net sown area of the state (Government of Punjab 2005). With the adoption of rice wheat cropping pattern in the state, crop diversity has decreased considerably and area under crops like, gram, pulses, groundnut, etc., which have a positive impact on soil quality, has decreased. Also, area under low input crops, like maize, *bajra, jowar* (sorghum), etc., have also decreased (Table 1.2).

Under the rice-wheat cropping pattern, rice has to be harvested early in order to accommodate the wheat crop. This means, a very little time is left in the hands of the farmers to turn around for planting the wheat crop. Within this period, the

---

[2] Statistical Abstract of Punjab, (Government of Punjab 2005).

Table 1.2 Cropping pattern in Punjab (GCA in thousand hectares) (Area under various crops as percentage of gross cropped area)

Crop	1960–1961	1970–1971	1980–1981	1990–1991	2000–2001	2006–2007	2010–2011	2011–2012
Rice	4.79	6.87	17.49	26.86	32.89	32.90	35.72	35.62
Wheat	29.58	40.49	41.58	43.63	42.92	44.00	44.36	44.59
Maize	6.91	9.77	4.50	2.44	2.08	1.94	1.68	1.59
Bajra and jowar	2.72	3.73	1.03	0.16	0.08	0.13	0.04	0.04
Cotton (American)	5.17	3.73	7.42	8.49	4.51	7.22	5.94	6.40
Cotton (Desi)	4.26	3.26	2.17	0.85	1.46	0.43	0.16	0.11
Sugarcane	2.81	2.25	1.05	1.35	1.52	1.39	0.88	1.01
Total oilseeds	3.90	3.96	3.52	1.39	1.08	0.81	0.49	0.47
Total pulses	19.08	7.29	5.04	1.91	0.68	0.36	0.27	0.23
Barley	1.39	1.00	0.96	0.49	0.40	0.24	0.15	0.15
Vegetables	–	0.88	0.95	0.72	1.39	1.39	1.40	1.40
Fruits	–	0.88	0.43	0.92	0.43	0.72	0.88	0.90
Other crops	19.39	15.89	13.86	10.79	10.56	8.47	8.02	7.49
Gross cropped area	4,732	5,678	6,763	7,502	7,941	7,932	7,912	7,912

Source Statistical abstracts of Punjab, various issues

farmer has to get rid of the rice stubble and prepare the land for sowing the wheat crop. The previous varieties of rice and wheat crops were of long duration and could fit rice-wheat rotation only in small areas. But with the availability of photo-period non sensitive short duration varieties of wheat as well as rice it became possible to grow high yielding 120–130 days rice crop, i.e., June–July to October–November followed by a high yielding 110–120 days wheat crop, i.e., November–December to March–April. With the adoption of these varieties rice-wheat crop rotation was practiced in areas which formerly produced only wheat or rice but not both in the same field in any one farming year. The major constraint in the rice-wheat cropping system is the available short time between rice harvesting and sowing of wheat and any delay in sowing adversely affects the wheat crop. Preparation of the field also involves removal or utilization of rice straw left in the field.

Various modern inputs were introduced in Punjab to harvest the rice crop within such a short period of time. One such input which has become the most popular implement in the rice-wheat cropping system is the use of the combined mechanized harvester. The use of the combined harvester has increased at a tremendous rate in Punjab. Almost 80 % of the rice crop is harvested using this implement in Punjab. However, the use of the combined harvester has in reality exacerbated the problem of crop residue management. The use of combined harvesters leaves behind a large amount of rice residue to be burnt in the open fields. The combined harvester spreads the rice residue in the fields which is difficult to collect. It is widely perceived that farmers find it the easiest and the most economical way of getting rid of the rice stubble through burning it. Also, the shortage of time for sowing the wheat crop, after the rice crop harvest, leaves farmers with no other option but to burn it.

Thus, burning has emerged as the standard method of rice residue/stubble management in the combine harvested rice-wheat cropping system that is practised on a broad scale in the state of Punjab in northwest India. Every year almost 15 million tonnes of paddy straw are generated in Punjab. Of this, according to various estimates, on an average, almost 7–8 million tonnes of rice residue are set on fire in open fields.

Rice residue burning results in extensive impacts both on and off farm, e.g., losses in soil nutrients, soil organic matter, production and productivity, air quality, biodiversity, and water and energy efficiency and on human and animal health. In India, air pollution from residue burning can be severe, with impacts on human health by directly causing or exacerbating a range of health hazards and contributing to the incidence of traumatic road accidents through significantly reduced visibility. One of the recognized threats to the rice-wheat cropping system sustainability is the loss of soil organic matter as a result of burning. The straw collected from the fields is of great economic value as livestock feed, fuel and industrial raw material. In northern India, wheat straw is preferred while in Southern India paddy straw is fed to livestock (Hegde 2010). The residue generated from the rice-wheat cropping system can be put to many uses, but this is possible if the residue is separated from the grain and carried out of the field. Burning reduces the availability of straw to livestock, which is already in short supply by more than 40 %. However, in the case of combine harvesting, most of the residue is left in the field for burning adversely affecting overall sustainability of the rice-wheat cropping system (Thakur 2003). Zero tillage after stubble burning is now being adopted by many farmers. In 2005–2006, around 10 % of the total area sown under wheat was by using zero till machines. Apparently less than 1 % of farmers incorporate the paddy straw because in the case of incorporation more tillage operations are required than after burning (Singh et al. 2008). The options for crop residue management may include developing systems to plant residue into bailing and removal for use as animal feed or for industry. Enhanced decomposition of machine-harvested straw to improve nutrients in the soil can be useful. The use of microbial sprays that can speed decomposition of residue is also an option. The option of planting into residue needs further investigation of inorganic nitrogen and its adverse effect due to nitrogen deficiency.

Though various studies in the literature have addressed this issue of burning of the crop stubble but none have brought to the forefront the adverse implications of this unwarranted practice on human and animal health. The study proceeds first by bringing to the forefront the amount of pollution being caused by rice residue burning. Thereafter the harmful effects of the pollution being generated by rice stubble burning on human health are studied. Based on the information obtained, we analyse the Punjab government's existing policies to address air pollution caused by rice stubble burning. What policies have the Punjab government put in place to prevent this practice? Are there any bottlenecks in the actual implementation of these policies? What are the current mechanisms in place for recording and monitoring the pollution caused by crop stubble burning? Based on the findings of the Punjab government policies to address the pollution caused by crop stubble burning, the study aims to provide policy suggestions to remove the practice.

The study aims to estimate the monitory value of health effect of crop stubble burning in rural Punjab. However, it needs to be highlighted here that crop stubble burning leads to various losses including loss in soil nutrients, soil organic matter, productivity of soil, water and energy efficiency in addition to its adverse impact on human and animal health and its impact on vegetation, air quality, environment and biodiversity. The subject matter of the present study only deals with the adverse impact of stubble burning on human health, which is measureable in monetary terms, e.g., the amount spent on treatment, medicine, cure and losses in working hours. The other losses mentioned above have not been attempted in this study and therefore such losses are beyond the subject matter of the present study and should be understood as a limitation of this study.

Based on the findings of the Australian Council of International Agriculture Research (ACIAR) Project, LWR/2006/124, 'Fine-tuning the Happy Seeder Technology for Adoption in the Northwest India' on the feasibility of the various alternatives to crop residue burning, and based on our own information collected from various departments of the Punjab government, the study analyses the viability of some alternative residue uses such as fodder for animals, fuel for the generation of electricity, etc. As part of the ACIAR project LWR/2000/089 (Permanent beds for irrigated rice-wheat and alternative cropping systems in northwest India and southeast Australia), a new generation of seeders capable of directly drilling wheat into heavy rice residue loads without prior burning was developed. These machines have been called Happy Seeders. Preliminary financial evaluation of the technology within LWR/2000/089 indicates that adoption of the technology can be both financially viable for farmers, and financially preferable to alternative residue management practices such as residue incorporation or residue burning. In addition, preliminary economic evaluation of important external benefits associated with the use of the Happy Seeder, such as reduced public costs in the provision of fertilizer and irrigation water to farmers, suggests that there may be substantial gains for the broader community from adoption of the Happy Seeder in the form of lower levels of air pollution.

Existing policy settings and/or the way they are practically interpreted and implemented may constrain the adoption of the Happy Seeder technology by farmers. The ACIAR project PLIA/2006/180 (Happy Seeder policy linkage scoping study) assessed the range and scale of policy related issues for its adoption. The scoping study identified a range of constraints and recommended that more measures thorough financial and economic evaluation of the technology and its alternatives, and the assessment of potential policy instruments which could be used to enhance adoption, be undertaken.

The ACIAR project proposal LWR/2006/124 (Fine tuning the "Happy Seeder" technology for adoption in northwest India) includes objectives to extend and further refine the financial evaluation of the Happy Seeder technology relative to options involving the burning or incorporation of rice residue. These financial evaluations are important in determining the viability and private incentive for adoption of the technology from the point of view of individual farmers, depending on farm size, cost structures, etc. However, these evaluations are not designed to inform policy interventions which may be necessary to enhance adoption to levels consistent with generating significant reductions in off-site impacts.

The present study aims to broaden the analysis beyond the farm and financial levels. It analyses off-site uses of rice residue; undertakes the analysis from a socio-economic rather than only financial perspective; For example, some proponents of the Happy Seeder currently favour enforcement of existing pollution laws which ban residue burning in combination with the provision of government subsidies to individual farmers to lower the initial capital cost of purchasing the Happy Seeder machinery. However, appropriate assessments have not been undertaken to demonstrate the relevant impact of agriculturally based pollution to broad-scale air pollution; the book addresses these questions.

This book is an outcome of the project carried out by the authors titled, 'Policy instruments to address air pollution issues in agriculture—Implications for Happy Seeder Technology Adoption in India'. The project was funded by the ACIAR and was carried out at the National Council of Applied Economic Research, New Delhi. The development of this project is in response to recommendations made within the ACIAR scoping study PLIA/2000/180 (Happy Seeder policy linkages scoping study). The scoping study identified a range of policy related constraints, in particular:

- An inadequate understanding of the financial viability of the technology over a range of farmers' socio-economic circumstances (often relating to farm size);
- The significant but unquantified external benefits that would accrue to the broader community from adoption of the Happy Seeder technology often relating to pollution reduction;
- A lack of analysis of the economic performance of the Happy Seeder technology relative to the performance of other off-site uses of rice residues; and,
- A focus by relevant state policy makers on the financial performance of the technology in their consideration of the need for government intervention to accelerate and increase adoption rather than focusing on both financial and economic performance.

As a consequence, the scoping study recommended that more comprehensive financial and socio-economic evaluation of the technology and its alternatives, and the assessment of a range of potential policy instruments which could be used to enhance adoption, be undertaken. The study offers a range of potential economic, social and environmental insights. Preliminary studies show that the Happy Seeder technology may offer potential economic benefits over traditional residue burning activities in the rice-wheat production system.[3] The Happy Seeder technology is relevant to a large area of the northwest Indo-Gangetic plains of India in which the rice-wheat production system predominates. It is proposed within the ACIAR project proposal LWR/2006/124 that if the Happy Seeder technology is utilized over 10 % of the area currently under zero till plus burning regime in Punjab, it would result in potential financial benefits of Rs. 92 million (approximately, A$2.7 million). Accounting for externalities would result in potential economic benefits of an even larger magnitude.

Overcoming impediments to the adoption of less polluting agricultural technologies will be of significant benefit to the broader community. Benefits will include reduction in a range of off-site impacts of residue burning, including those on human health, other industries, and adjacent communities, especially smoke-related illness, transport disruption, etc. Social benefits will accrue from designing government adoption incentives which better account for the range of socio-economic circumstances of rice-wheat farmers in Punjab.

Along with reduction in air and water pollution, higher levels of adoption of less polluting agricultural technologies ensure improvements in soil health, primarily through improvements in soil nutrient levels and soil organic matter, and reductions in the irrigation water and electricity demands for groundwater pumping in the rice-wheat production system. Biodiversity conservation also gets enhanced through a decline in residue burning as it reduces fire damage to adjacent remnant vegetation and wildlife habitat including nationally significant species.

The analysis gauges the relative significance of policies and other drivers of changes in residue management practices, research and/or development strategy. The study aims to resolve policy issues identified in the ACIAR project PLIA/2006/180 (Happy Seeder policy linkage scoping study). The Happy Seeder technology was developed and proof of concept achieved in the ACIAR project LWR/2000/089 (Permanent beds for irrigated rice-wheat and alternative cropping systems in northwest India and southeast Australia).

The study targets air pollution issues in agriculture within the state of Punjab. Its findings and recommendations will be useful and relevant to policy makers and analysts from organizations such as the Punjab Pollution Control Board, Punjab State Department of Agriculture, and Punjab State Council for Science and Technology. The Punjab government has recently established a government taskforce on air pollution/residue burning, chaired by the Director of the Punjab Pollution Control Board. The taskforce is keenly interested in the findings of the

---

[3] The results of the financial analysis indicate that the present value of total financial benefits from adoption of the Happy Seeder in comparison to residue burnt/zero tillage wheat are Rs. 10,150/ha higher (A$299), and Rs. 32,750/ha (A$963/ha) higher than the residue burnt/conventional tillage option. When economic values are incorporated for the water use efficiency related cost savings, the benefits of the Happy Seeder are further increased to Rs. 20,000/ha (A$588/ha) and Rs. 42,500/ha (A$1,250/ha) over these other options, respectively (Sidhu et al. 2007).

study, as the mandate of the taskforce is to draft a policy on residue burning and pollution issues in Punjab agriculture.

Results of the Australian component of the project directly communicate to policy makers within the NSW Department of Primary Industries (DPI) and the NSW Department of Environment and Climate Change (DECC). These agencies have primary interests and responsibilities for the development of sustainable practices and natural resource management policies. The findings and recommendations would be relevant to manage crop residue concerns for rice, with most of the Australian rice industry being located in NSW, and other crops.

1.3 Main Objectives of the Study

The study focuses on environmental policy issues associated with rice residue burning. The main objectives of the study are:

1. Assess the broader significance of agriculture-based pollution in Punjab and describe existing and proposed policies.
2. Evaluate the cost of pollution caused by agricultural waste burning to the society at large and suggest a range of potential alternate uses of rice/wheat stubble.
3. Review the relative significance of policies and technologies in changing residue management practices in Punjab.

Assessment undertaken in addressing above objectives was completed though the review of existing secondary data, reports and publications and through interviews and discussions with concerned policy makers and analysts. For the assessment of economic valuation of pollution on human health, a primary household survey of 150 households was carried out in three villages in the Patiala district. The details of methodology are provided in each of the chapters. The review of the technology context associated with historical reductions in burning as a residue management practice in India has been undertaken through the review of reports and publications relating to technological change in this area, and of its adoption by Indian farmers. Review of the policy context was done through analyses of historical legislative changes and government programmes and incentives, available published and unpublished relevant documents, and through interviews and discussions with Punjab government officials.

1.4 An Overview

The book includes seven chapters including the introduction to the study. This chapter also lays down the details of genesis of this study and puts forth the main objectives for which the study was undertaken, and the methodology followed and data used in meeting with those objectives. Chapter 2 gives an overview of management of crop stubble. The chapter introduces the extent of pollution caused by crop stubble burning, citing various examples from the literature, its effect on fertility of soil, impact on human health and available different methods of crop stubble management as cited in the literature and alternative uses of crop stubble. Chapter 3 deals with evaluating the health effects of air pollution from agricultural residue burning. This chapter presents the details of the design of the household survey in the three selected study villages followed by presentation of ambient air quality levels in the villages during the period when harvesting of rice take place. Subsequently, some details of the agricultural output (productivity) among the selected households and some health indicators are presented in the chapter followed by theoretical model and estimation strategy and the results on monetary estimates of health consequences of air pollution.

Chapter 4 looks at the alternative uses of crop stubble. The chapter starts with presenting disposal pattern of paddy straw giving details of alternate uses of agriculture waste, viz., rice residue as fodder for animals, its use in bio-thermal power plants, its use for bedding material for animals, mushroom cultivation and so on. The chapter discusses in details of residue use in power generation citing various bio-mass power projects commissioned in the state by the Punjab Energy Development Agency (PEDA).

Chapters 5 and 6 present details of environment legislation in the state as well as the country as a whole and the Punjab government's policy to tackle the problem of agricultural waste burning. Chapter 5 discusses the legislation on pollution in India in general and Punjab in particular. The chapter presents various laws to control pollution like Water Act 1974, Air Prevention and Control of Pollution Act 1981, Environment Protection Act 1986, National Environment Tribunal Act 1995, Noise Pollution Rules 2000, Bio-diversity Act 2000 and so on. The chapter also discusses various functions and activities of Central Pollution Control Board, Punjab Pollution Control Board, Punjab State Council for Science and Technology, Punjab Energy Development Agency and Punjab Bio Diversity Board. Chapter 6 analyses Punjab government policies for restricting the agriculture residue burning. The chapter discusses the role being played by different organs of Punjab government in controlling agriculture waste burning, especially to mention, Punjab Pollution Control Board, Agricultural Councils, Punjab State Council for Science and Technology, Department of Agriculture, Department of Animal Husbandry, Punjab Energy Development Agency, Punjab Agricultural University, Punjab State Farmers' Commission, Department of Rural Development and *panchayat*s (local governments). Chapter 7 offers summary conclusions and policy suggestions.

References

Benbi, D. K., Nayyar, V. K., & Brar, J. S. (2006). The green revolution in Punjab: Impact on soil health. *Indian Journal of Fertilizers,2*(4), 57–66.

Government of India. (2011). *Dynamic groundwater resources of India.* Faridabad: Central Ground Water Board, Ministry of Water Resources, Government of India.

Government of Punjab. (2005). Economic and Statistical Organization, Statistical Abstract of Punjab.

Hegde N. G. (2010). Forage resource development in India. In: *Souvenir of IGFRI Foundation Day*, November, 2010. www.baif.org.in.

Singh, J., & Hussain, M. (2002). Total factor productivity analysis and its components in a high potential rice-wheat system: A case study of the Indian Punjab. In M. Sombilla, M. Hussain, & B. Hardy (Eds.), *Developments in Asian rice economy.* Manila: IRRI.

Sidhu, H. S., Singh, M., Humphreys, E., Singh, Y., Singh, B., Dhillon, S. S., (2007). The Happy Seeder enables direct drilling of wheat into rice stubble. *Australian Journal of Experimental Agriculture,47*(7), 844–854.

Singh RP, Dhaliwal HS, Humphreys E, Sidhu HS, Singh M, Singh Y, John B (2008) *Economic Assessment of the Happy Seeder for Rice-Wheat Systems in Punjab, India.* A paper presented at AARES 52nd annual conference, Canberra, ACT, Australia.

Thakur, T. C. (2003). Crop residue as animal feed: Addressing resource conservation issues in rice–wheat systems of south Asia, a resource book. Rice Wheat Consortium for Indo-Gangetic Plains (CIMMYT), March, 2003.

Chapter 2
The Extent and Management of Crop Stubble

Abstract Burning of farm waste causes severe pollution of land and water on local as well as regional scales. It is estimated that burning of paddy straw results in nutrient losses viz., 3.85 million tonnes of organic carbon, 59,000 t of nitrogen, 20,000 t of phosphorus and 34,000 t of potassium. This also adversely affects the nutrient budget in the soil. It results in the emission of smoke which if added to the gases present in the air like methane, nitrogen oxide and ammonia, can cause severe atmospheric pollution. These gaseous emissions can result in health risk, aggravating asthma, chronic bronchitis and decreased lung function. Burning of crop residue also contributes indirectly to the increased ozone pollution. The chapter puts forth literature on various aspects of residue generated on the field, chemical composition of the residue, volume of pollution caused by residue burning, adverse impact of burning on human and animal health and various ways of crop stubble management.

Keywords Crop stubble burning · Chemical composition of residue · Health impact of stubble burning · Stubble burning and soil fertility · Crop stubble management

2.1 Introduction

Production and consumption activities generate pollution and waste, and atmospheric environment can absorb pollution/waste up to a limit. Agriculture is one of the important production activities and crop residue burning generates a significant amount of air pollution. Atmospheric environment can absorb this pollution in a particular geographic region given its assimilative capacity. If the burning activities remain confined within the assimilative capacity, the pollution does not create harmful effects. Therefore, in the initial stages when the production and burning activities are limited, pollution caused through these activities is not considered a problem. However, due to technological advancements in the agricultural sector,

© The Author(s) 2015
P. Kumar et al., *Socioeconomic and Environmental Implications of Agricultural Residue Burning*, SpringerBriefs in Environmental Science,
DOI 10.1007/978-81-322-2014-5_2

waste concentration has gone beyond the assimilative capacity of the environmental limit, thereby distorting the balance.

Burning of farm waste causes severe pollution of land and water on local as well as regional and global scales. It is estimated that burning of paddy straw results in annual nutrient losses to the tune of 3.85 million tonnes of organic carbon, 59,000 t of nitrogen, 20,000 t of phosphorus and 34,000 t of potassium at the aggregate. This also adversely affects the nutrient composition of the soil. When crop residue is burnt existing minerals present in the soil are destroyed, which adversely hampers the cultivation of the next crop. Straw carbon, nitrogen and sulphur are completely burnt and lost in the atmosphere in the process of burning. This results in the emission of smoke which when added to the gases present in the air like methane, nitrogen oxide and ammonia can cause severe atmospheric pollution. These gaseous emissions can pose health risks, aggravating asthma, chronic bronchitis and decreased lung function. Burning of crop residue also contributes indirectly to increased ozone pollution.

The chapter is organized as follows: the next section introduces the amount of crop stubble produced and the extent of this being burnt in the field. Section 2.3 presents the pollution caused by crop stubble burning, citing various discussions from the literature, followed by a section on the effects of crop stubble burning on soil fertility. Section 2.5 concentrates on the health impacts of pollution due to residue burning. The last section presents the management of crop stubble, like in situ, alternate uses of crop stubble, cost of alternate uses and end use of rice residue.

2.2 The Produce of Crop Stubble and Its Burning

Various studies have brought to the forefront the quantity of crop stubble generated in India and the proportion of wheat and rice stubble in the total crop stubble (Table 2.1). As per different studies, the residues of rice and wheat crops are major contributors in the total stubble loads in India. One such study by Garg (2008) estimates the contribution of rice and wheat stubble loads in the total stubble as 36 and 41 %, respectively in the year 2000, while the contribution of Punjab in the total burnt stubble of rice and wheat to be 11 and 36 %, respectively during the same time period. Table 2.2 provides the estimates of residue management practices followed in Punjab.

According to Mandal et al. (2004), the total amount of crop residue generated in India is estimated at 350×10^6 kg year^{-1} of which wheat residue constitutes about 27 % and rice residue about 51 %. According to Gupta et al. (2004), the total crop residue produced in India during 2000 was 347 million tonnes, of which rice and wheat crop residues together constituted more than 200 million tonnes.

According to Sidhu and Beri (2005), total production of paddy stubble in Punjab in 2004–2005 reached 18.8 million tonnes, of which 15 million tonnes

Table 2.1 Total quantity of crop stubble generated in India as per different studies

Study and year	Total quantity of crop residue produced in India
Garg (2008)	133,138 Gg
Mandal et al. (2004)	350×10^6 kg year^{-1}
Gupta et al. (2004)	347 million tonnes (2000)
Agarwal et al. (2008)	184,902 Gg

Table 2.2 End use of stubble by the farmers

End use	Rice (percentage of total stubble production)	Wheat (percentage of total stubble production)
Fodder	7	45
Soil incorporation	1	<1
Burnt	81	48
Rope making	4	0
Miscellaneous	7	7

Source Government of Punjab (2007)

were burnt in open fields. The study further quotes that 80 % of the rice harvested using combined harvester is burnt in open fields. However, according to Singh et al. (2008), around 17 million tonnes of paddy straw are produced every year in Punjab, of which 90 % are burnt in open fields.

Another study by Punia et al. (2008) attempts to estimate district-wise burnt area of agricultural residue by using remote sensing data. The total stubble burnt on the area in Punjab was found to be around 4,315.35 km^2 as on 15 May 2005. Among these, Amritsar had 673.99 km^2 of burnt area followed by Jalandhar, Ludhiana, Firozpur and Patiala districts while Roop Nagar had the least burnt area (41.36 km^2). During the field visit to Punjab, it was discovered that burning is done in two ways. One is partial burning which involves running of combine harvester followed by burning of small stalks and another of complete burning in which the entire field is set on fire while the latter process is mostly practiced. Table 2.3 provides a summary of various studies reporting rice residue burning practices in open field in Punjab. Both practices cause pollution but the impact is more severe in the case of complete burning. The farmers in the region resort to burning of straw due to lack of available economically viable options for managing the residue.

To conclude, although there is no unanimity in analysing the extent of residue generated and the percentage burnt in open fields. However, most of the studies in the literature are in agreement that as high as around two-thirds to three-fourths of the residue are being burnt in the case of paddy, mainly because of uneconomical options available to farmers for any alternate use of the same.

Table 2.3 Various studies reporting rice residue burnt in open fields in Punjab

Literature and year of study	Percentage of rice and wheat crop harvested using a combined harvester in Punjab	Dry fodder yield per hectare or aggregate production in Punjab		Quantity of rice residue burnt in open fields in Punjab
		Rice	Wheat	
Badarinath and Chand Kiran (2006)	75–80 %	6.2–11.8 t ha^{-1}	3.2–5.6 t ha^{-1}	70–80 million tonnes
Mandal et al. (2004)	More than 75 %	Data not available	Data not available	10 t ha^{-1}
Sidhu et al. (1998)	88.6 % for rice 56.6 % for wheat	Data not available	Data not available	Varies according to different districts, the highest being for Bhatinda where all the rice stubble is burnt, followed by Faridkot at 97.6 %, Ludhiana and Sangrur at 95 %, each
Gupta et al. (2004)	75 %	78 million tonnes (2000)	85 million tonnes (2000)	Data not available
Sidhu and Beri (2005)	More than 75 %	Data not available	Data not available	80 % of the total harvested using combined harvesters
Badve (1991)	Data not available	134.35 million tonnes (1983–1984)	67.71 million tonnes (1983–1984)	Data not available
Venkataraman et al. (2006)	Data not available	Data not available	Data not available	30–40 % straw burnt in Indo-Gangetic Plain (IGP)
Sarkar et al. (1999)	75 % combine harvested	Data not available	Data not available	100 % combined harvested burnt

Table 2.4 Residue to product ratio according to various studies

References	RPR ratio/quantity	Reason
AIT-EEC (1983)	0.416	If only the top portion of the rice stem along with 3–5 leaves is cut, leaving the remaining in the field
Bhattacharya and Shrestha (1990)	0.452	
Bhattacharya et al. (1993)	1.757	When the rice is cut at about 2 inches from the ground
Vimal (1979)	1.875	
Sidhu et al. (1998)	1.5:1	For both the crops
Gadde et al. (2009)	0.75	
Gupta et al. (2004)	1.5:1	Every 4 t of rice or wheat grain produces about 6 t of rice or wheat straw
Njie (2006)	1.25	For rice
Bhattacharya et al. (1993)	1.76	For rice
Singh and Rangnekar (1986)	1.5:1	For both rice and wheat
Koopman and Koppejan (1997)	1.757 and 0.267	For paddy straw and rice husk respectively
Badarinath and Chand Kiran (2006)	(3.2–5.6 t/ha and 6.2–11.8 t /ha)	For wheat and rice respectively
Sidhu and Beri (2005)	18.75 Mt	Of rice residue in the entire Punjab

2.2.1 Straw/Residue to Grain Ratio

To obtain the average amount of paddy straw generated and burnt in the state, the Residue to Product Ratio (RPR) must be known. Different studies on the subject of crop stubble in India have considered different residue/stubble to product ratio (RPR). As per these studies, the residue to product ratio (RPR) varies from 0.416 to 3.96. Table 2.4 provides the estimates of RPR obtained in various studies. The variation in the residue to product ratio is due to the overtime improvements in technology in the form of HYVs, irrigation, fertilizers, etc.[1] While estimating the quantity of straw produced, basmati rice varieties were not considered purposely as harvesting of basmati paddy is cultivated manually. Therefore, no straw is left behind for open burning in the field in the case of basmati. The principal reason for not having the practice of basmati straw being burnt is its use as feed for the animals. This also indicates that if the technology develops, making paddy straw silica free and nutritious for serving it as feed for the animals, farmers may stop burning the same, although it would also require an economical bailing system for its post combine harvest collection.

---

[1] Improvement in technology alone cannot explain such a large variation in RPR.

2.2.2 Chemical Composition of Rice and Wheat Stubble

Crop residue is not a waste but rather a useful natural resource. About 25 % of nitrogen (N) and phosphorus (P), 50 % of sulphur (S) and 75 % of potassium (K) uptake by cereal crops are retained in crop residues, making them valuable nutrient sources. Sidhu et al. (2007) estimated the quantity of nutrients available in rice. According to his study, the paddy straw has 39 kg/ha N, 6 kg/ha P, 140 kg/ha K and 11 kg/ha S. When transformed into monetary values it becomes Rs. 424 ha^{-1} of N, Rs. 96 ha^{-1} of P and Rs. 231 ha^{-1} of S, i.e., a sum equal to Rs. 751 ha^{-1}.

Sidhu and Beri (2005) shared their experience with managing rice residues in intensive rice-wheat cropping system in Punjab. According to them, the approximate amount of the nutrients present in the straw, which was burnt in 2003–2004 were 106, 65 and 237 thousand tonnes respectively of N, P_2O_5 and K_2O in addition to secondary and micronutrients. Tables 2.5, 2.6, 2.7 and 2.8 show the nutrient content of rice and wheat straw and amounts removed with 1 t of straw residue and their chemical compositions.[2]

Table 2.5 Nutrient content of paddy straw and amounts removed with one tonne of straw residue

	N	P_2O_5	K_2O	S	Si
Content in straw, percent dry matter	0.5–0.8	0.16–0.27	1.4–2.0	0.05–0.10	4–7
Removal with 1 t straw, kg/ha	5–8	1.6–2.7	14–20	0.5–1.0	40–70

Source 'Rice straw management'; Dobermann and Fairhurst (2002)

Table 2.6 Nutrient content in rice residue

Rice	N	P_2O_5	K_2O
Nutrient Content in %	0.61	0.18	1.38

Source Mandal et al. (2004)

Table 2.7 Moisture and other factors in rice residue

Literature study	Moisture content in percent	C percent	N percent	LHV (lower heating value) MJ/kg	Ash percent
AIT-EEC (1983)	27.00	–	–	15.10	16.98
Strehler and Stutzle (1987)	12–22	41.44	0.67	10.90	17.40
Bhattacharya and Shrestha (1990)	12.71	24.79	–	16.02	21.05
Bhattacharya et al. (1993)	12.71	39.84	–	16.02	–

[2] The nutrient contents will vary as per the agro-biological and soil conditions.

Table 2.8 Chemical composition in rice and wheat straw

Name of the crop	Organic matter	Crude protein	Crude fibre	Ash
	Percentage composition			
Rice straw	82.0	4.0	37.0	18.0
Wheat straw	–	3.5	–	7.5

Source Agarwal et al. (2008)

2.3 Volume of Pollution Caused by Crop Stubble Burning

Open field burning of crop stubble results in the emission of many harmful gases in the atmosphere, like carbon monoxide, N_2O, NO_2, SO_2, CH_4 along with particulate matter and hydrocarbons. These trace gases have adverse implications not only on the atmosphere but also on human and animal health (Gupta and Sahai 2005; Lal 2006; Agarwal et al. 2006; Canadian Lung Association 2007). These also result in the loss of plant nutrients and thus adversely affect soil properties. It has been estimated that for the year 2000, the emission of CH_4, CO, N_2O and NO_2 was 110, 2306, 2 and 84 Gg respectively, from the field-burning of rice and wheat straw (Mandal et al. 2004).

A study conducted by the National Remote Sensing Agency in Punjab reported that wheat crop residue burning contributed about 113 Gg (Giga gram = 10 billion gram) of CO, 8.6 Gg of NO_2, 1.33 Gg of CH_4, 13 Gg of PM_{10} and 12 Gg of $PM_{2.5}$ during May 2005 and paddy straw/stubble burning was estimated to contribute 261 Gg of CO, 19.8 Gg of NO_2, 3 Gh of CH_4, 30 Gg of PM_{10} and 28.3 Gg of $PM_{2.5}$ during October 2005 (Badarinath and Chand Kiran 2006).

The information provided by Punjab Agricultural University (PAU) to the State Environmental Council also estimated that the crop residues contained about 6.0 million tonnes of carbon that on burning could produce about 22.0 million tonnes of carbon dioxide in a short span of 15–20 days. Additionally, the smoke fumes contain particulates of partially combusted materials as soot, which become airborne and are transported downwind, especially during winters when inversion sets in. Studies conducted by the Punjab Pollution Control Board in 2006 in villages, namely, Dhanouri, Simbro and Ajnouda Kalan in the district of Patiala also indicated that CO and pollutant particulates were of major concern (PPCB 2007). CO appeared to be most critical as concentrations of 114.5 mg/m^3 or more were observed at 30 m distance from burning fields and 20.6 mg/m^3 CO was recorded at residences even 150 m away. Given that the permissible limit of CO in ambient air is 4.0 µg/m^3, this was a major health hazard for residents and road travellers in the area. Further, particulates were also being released in large quantities. $PM_{2.5}$ ranged between 146 and 221 µg/m^3 in critically affected areas and average PM_{10} values were found 300 µg/m^3, against a permissible limit of 60 µg/m^3 for residential rural areas. Significant amounts (40–50 µg/m^3) of NO_2 and NH_3 were also recorded during burning, at residences located 200–4,000 m away from burning site, though concentration of SO_2 was less. Further, concentrations of organic pollutants were also found to be significantly high. The smoke was also found to be toxic due to presence of heavy metals, especially iron and zinc. Iron concentrations were in the range of 6,778–13,240 $\mu g/m^3$, whereas zinc concentrations were in the range of 1,021–4,854 $\mu g/m^3$.

One tonne of straw on burning releases 3 kg of particulate matter, 60 kg of CO, 1,460 kg of CO_2, 199 kg of ash and 2 kg of SO_2 (Gupta et al. 2004). According to Singh et al. (2008), the major pollutants that are emitted during crop residue burning are given in Table 2.9. Further, in the year 2000, around 78 million tonnes of rice and 85 million tonnes of wheat straw were generated in India of which around 17–18 million tonnes ended up being burnt in the field.

Badrinath et al. (2008) used Indian Remote Sensing Satellite (IRS-P6) Advanced Wide Field Sensor (AWFS) data during May and October 2005 for estimating the extent of burnt areas and the resulting Green House Emissions (GHG) from crop residue burning. The authors found that the emissions from wheat residue in Punjab were relatively low as compared to the paddy residue. Roughly around 75–80 % of the rice was machine harvested, leaving behind large quantities of organic matter.

Venkataraman et al. (2006) calculated the state-wise crop waste generated using crop production data for 13 different crops for India. The waste to grain ratio or residue to product ratio was used as reported in various literatures such as Koopman and Koppejan (1997), Singh and Rangnekar (1986), and Bhattacharya et al. (1993) to calculate the waste generated. Dry matter fraction and combustion efficiency specific to crop waste types were used as reported in Koopman and Koppejan (1997), Smill (1999) and Streets et al. (2003). The study finds that the estimated crop waste generated was largely from cereal straws, with the waste generation being highest in north and western India. Most of the crop land fires took place in the western Indo-Gangetic plain which reaches its peak in May and October, the harvest season of *kharif* (sown in the monsoon season) and *rabi*

Table 2.9 Major pollutants emitted during crop residue burning

Category	Pollutants	Source
Particulars	SPM (PM_{100})	Incomplete combustion of in organic material, particle on burnt soil
	RPM (PM_{10})	Condensation after combustion of gases and incomplete combustion of organic matter
	FPM ($PM_{2.5}$)	
Gases	CO	Incomplete combustion of organic matter
	NO_2	Oxidation of N_2 in air at high temperature
	N_2O	
	O_3	Secondary pollutant, form due to Nitrogen Oxide and Hydrocarbon
	CH_4/Benzen	Incomplete combustion of organic matter
	PAH_S	Incomplete combustion of organic matter

SPM Suspended particulate matter; *PM* particulate matter; *FPM* fine particulate matter
Source Singh et al. (2008)

(sown in winter) crops, respectively. On an all-India basis, 18–30 % of the crop waste is burnt, but for the Gangetic plain the figure is much higher at 30–40 %. According to the findings of the study, all crop waste is burnt in the field in the states of Punjab, Uttar Pradesh and Haryana. In all these states not only the unu-tilized cereals namely, rice, wheat, barley, etc., and sugarcane waste are burnt in the field but waste from oilseeds, fibre crops, and pulses are also burnt. The study assumes that cereal straws and sugarcane straws are completely burnt in the field.

Venkataraman et al. (2006) found that the percentage of crop stubble in the animal feed was based on the estimated roughage in the diet and ranged from 74 to 85 % for dairy and non-dairy cattle, 50–60 % for pigs and a minor 0–5 % for sheep and goats. The state-wise cattle and livestock population for 2000–2001 in the four major states was obtained from the 1992 cattle census. Animal fodder was found mainly from the cereal straws at 85 % with the balance from pulses and oilseeds.

Another study by Sidhu and Beri (2005) focuses on the impact of wheat yield through different rice residue management practices for an 11-year period from 1993–1994 to 2003–2004. The study shows that the yield of wheat crop is 0.50 t/ha higher on an average for all the 11 years, if the rice residue is incorporated in the soil 2–3 weeks before sowing wheat than if it is burnt. This implies that there are tradeoffs, and it seems the private benefits from burning outweigh the costs associated with non-burning; implying a need for social benefit cost analysis. There are tradeoffs, and it seems the private benefits from burning outweigh the costs associated with non-burning. A social analysis may generate opposite results.

The study by Butchaiah et al. (2009), estimated a total of 22,289 Gg of paddy straw surplus production in India of which 13,915 Gg was estimated to be burnt in the field. The study makes use of the proportion of the paddy straw subject to open burning, rough rice production and straw to grain ratio to arrive at the above figure. An RPR of 0.75 was used in the study. The production of rough rice was obtained by multiplying the rice production data by a factor 1.5. The rice pro-duction data was calculated as an average of 6 year period from 1999–2000 to 2004–2005. According to the study, the amount of rice stubble burnt in the field was calculated as per the surplus rice stubble left in the fields. According to the National Biomass Resource Assessment (NBRA), 23 % of the total amount of paddy straw produced in the field is in surplus. The states of Punjab and Haryana contribute 48 % and Uttar Pradesh 14 % to the total that is subject to open field burning. This study estimates 13.92 Tg quantity of paddy straw burnt in the open fields. Furthermore, this study estimates the emission of different harmful gases and particulate matter in the air using emission factors of different gases and par-ticulate matter in g/kg, combustion factor, which is assumed to be 0.80 for all the gases and multiplying this with the quantity of rice stubble burnt in the open fields to obtain the estimates.

Sidhu et al. (1998) undertook a survey covering 11 districts in Punjab survey-ing a total of 237 farmers. The results of the study indicate that about 99.5 % area of the surveyed farmers was irrigated. Farmers did not have accurate informa-tion on the quantity of rice and wheat straw produced. However information on the yields of crops in the previous years obtained from the farmers was used to calculate the quantity of rice and wheat stubble produced. Only 6 % of the farmers owned a combine but a large number of farmers used combines for rice and wheat crop harvesting and 14 % of the farmers used shredders for cutting the stubble after combine harvesting the rice. Out of the total area owned by the farmers, 75 % was under rice and 80.7 % was under the wheat crop. Out of total cropped area, 88.6 % of the rice and 56.6 % of wheat in 11 districts covered in the survey were harvested using combine harvesters and reapers. The highest area harvested by combine was around 99.4 % of rice in Ferozepore and 65.4 % of wheat in Jalandhar. Total quantity of the paddy straw for the surveyed farmers ranged from 43.8 t farmer^{-1} in Jalandhar to 73.1 t farmer^{-1} in Patiala district with an average of 59.2 t farmers^{-1}. The quantity of wheat straw ranged from 42.3 t farmer^{-1} in Gurdaspur to 64.6 t farmer^{-1} in Amritsar. The average land holding of the surveyed farmers in all the 11 districts was 10.26 ha with the highest average land holding being in Gurdaspur at 12.3 ha.

Out of all the 11 districts surveyed, the quantity of rice and wheat residue was highest in Sangrur at 1942.1 and 1,544.6 t, respectively followed by Patiala and Amritsar at 1,656.8 and 1,532.3 t, respectively for rice and Amritsar and Ludhiana districts at 1,291.1 and 1,205.6 t for wheat, respectively. Based on the study, on an average, for all the 11 districts surveyed in the study, 90 % of the rice stubble and 48 % of the wheat stubble was burnt. The study used the rice and wheat yields in the year 1995–1996 to calculate the proportion of rice and wheat stubble burnt and also the annual loss of N through burning of straw. Assuming that both rice and wheat straw contained 0.5 % N, the total N lost was estimated to be 85,506 t year^{-1}.

Open burning contributed to 25 % of black carbon, organic matter and carbon monoxide emissions, 9–13 % to $PM_{2.5}$ and CO_2 emissions and 1 % to SO_2 emissions (Venkataraman et al. 2006). Table 2.10 gives the estimates of biomass burned and emission of Aerosols and Trace gases for crop waste open burning. The crop waste burned in the fields range from 18 to 30 % and has strong regional variations.

A study by Gupta et al. (2004), estimated the emission of the following trace gases from the burning of rice and wheat residues in the years 1994 and 2000 as given in Table 2.11. Similarly an another study by Badrinath et al. (2008) indicates that nearly 5,504 km^2 of wheat crop area was burnt during May 2005, with the average biomass in the field after harvesting at about 5.94 t ha^{-1}. While for paddy about 12,685 km^2 of area was burnt during that period. The results of the study are summarized in Table 2.12.

According to Gadde et al. (2009), the annual contribution from crop residue burning in Asia is calculated to be 0.10 Tg of SO_2, 0.96 Tg of NO_2, 379 Tg of CO_2, 23 Tg of CO and 0.68 Tg of CH_4. Gupta et al. (2004) indicated that burning of straw also emits large amount of particulates that are composed of a wide variety of organic and inorganic species. One tonne straw on burning releases 3 particulate matter, 60 kg CO, 1,460 kg CO_2, 199 kg ash and 2 kg SO_2. These gases and aerosols consisting of carbonaceous matter have an important role to play in

Table 2.10 National estimates of biomass burned and emission of aerosols and trace gases for crop waste open burning

Pollutants	Crop waste burning (Emission factors Gg year^{-1})				
	Cereals	Sugarcane	Others	Total crop waste	Total open burning
Biomass Burned Tg year^{-1}	67–189	32–70	17–30	116–289	148–350
BC	55–292	19–49	12–31	86–372	102–409
OC	134–770	48–122	39–79	211–970	399–1,529
OM	287–1,250	97–247	60–143	444–1,639	663–2,303
PM$_{2.5}$	369–1,913	125–289	78–191	572–2,393	851–3,317
CO$_2$ Tg year^{-1}	102–353	48–131	25–55	175–539	224–638
CO, Tg year^{-1}	6–49	3–18	2–8	10–74	13–81
SO$_2$	27–113	13–42	7–18	46–172	66–238
NO$_X$	168–845	80–313	42–132	289–1,290	393–1,540
CH$_4$	181–762	86–283	45–119	313–1,164	420–1,486
NMVOC	1,055–4,430	500–1,644	263–693	1,818–6,767	2,039–7,406
NH$_3$	87–367	41–136	22–57	151–560	189–661

Source Venkataraman et al. (2006)

Table 2.11 Emission of trace gases from burning of rice and wheat residue

Year	CH$_4$	CO	N$_2$O	NO$_X$
1994	102	2138	2.2	78
2000	110	2,305	2.3	84

Source Gupta et al. (2004)

Table 2.12 Total emission by burning of rice and wheat crop

Name of the crop	Total Emissions Gg				
	CO	NO$_X$	CH$_4$	PM$_{10}$	PM$_{2.5}$
Wheat	113	8.6	1.33	13	12
Rice	261	19.8	3	30	28.3

Source Badarinath et al. (2008)

the atmospheric chemistry and can affect regional environment that also has linkages with global climate change.

Gupta et al. (2004) in their study estimated that burning of straw raises the soil temperature up to 33.8–42.2 °C (1-cm depth). About 23–73 % of nitrogen is lost and the fungal and bacterial population are decreased immediately up to 2.5 cm depth of soil. According to a study conducted by the Department of Soils, PAU, 2006, Punjab produces around 23 million tonnes of paddy straw and 17 million tonnes of wheat straw, annually. The burning of straw raises the temperature of the soil in the top 3 inches to such a high degree that the carbon-nitrogen equilibrium in soil changes rapidly. The carbon as CO$_2$ is lost to atmosphere, while nitrogen is converted to nitrate. This leads to a loss of about 0.824 million tonnes of NPK from the soil.

The above discussed studies clearly establish that mass agricultural residue burning in the fields is seriously damaging the environment. Further, open burning of residue in the fields also leads to death of soil micro flora and fauna and may also damage nearby trees in addition to adjoining standing crops. Further, the ash left after burning is a very good absorbent and if not fixed properly, absorbs the applied weedicides, which results in decreased efficacy of herbicides.

Thus, the on-site impact of burning includes removal of a large portion of the organic material, denying the soil an opportunity to enhance its organic matter and incorporate important chemicals such as nitrogen and phosphorus, as well as, loss of useful micro flora and fauna. The off-site impacts are health related due to general air quality degradation of the region resulting in aggravation of respiratory like cough, asthma, bronchitis, eye and skin diseases. Fine particles also can aggravate chronic heart and lung diseases and have been linked to premature deaths in people already suffering from these diseases. The black soot generated during burning also results in poor visibility which could lead to increased road side incidence of accidents. It is thus essential to mitigate impacts due to the burning of agricultural waste in the open fields and its consequent effects on soil, ambient air and living organisms.

2.4 Effects of Crop Stubble Burning on the Fertility of the Soil

As per the Department of Agriculture of the Punjab government, the soils of Punjab are generally low in nitrogen content, low to medium in phosphorus and medium to high in potassium. The soil organic carbon in Punjab has been reduced to very low and inadequate levels due to the inadequate application of organic manures and non-recycling of crop residues. A rice-wheat sequence that yields 7 t ha^{-1} of rice and 4 t ha^{-1} of wheat removes more than 300 kg of nitrogen, 30 kg of phosphorus and 300 kg of potassium per hectare from the soil. Another study estimates that a 10 t ha^{-1} crop yield removes 730 kg NPK from the soil.

The burning of crop stubble in open fields has adverse impact on the fertility of the soil, eroding the amount of nutrients present in the soil. Burning also kills soil borne deleterious pests and pathogens. According to the Department of Soil, Punjab Agricultural University, burning results in the soil organic carbon being lost to the atmosphere as CO_2, nitrogen equilibrium in the soil changes rapidly and nitrogen is converted to nitrate. As a result, there is a loss of 0.824 million tonnes of NPK from the soil.

According to Gupta et al. (2004), burning of crop stubble increases the temperature in the soil up to 33.8–42.2 °C. Burning also results in the loss of 27–73 % of nitrogen present in the soil and reduces the bacterial and fungal populations on the top 2.5 cm of the soil. Furthermore, repeated burning can diminish the bacterial population by more than 50 %. Long-term burning also reduces total nitrogen and carbon and potentially mineralized nitrogen in the 0–15 cm soil layer along with a loss in the soil organic matter.

Table 2.13   Nutrient losses due to burning of rice residues in Punjab, 2001–2002

Nutrient	Concentration in straw (g/kg)	Percentage lost in burning	Loss (kg/ha)
C	400	100	2,400
N	6.5	90	35
P	2.1	25	3.2
K	17.5	20	21
S	0.75	60	2.7

Source Singh et al. (2008)

As per Mandal et al. (2004), burning of rice and wheat residue results in the loss of about 80 % of nitrogen, 25 % of phosphorus, 21 % of potassium and 4–60 % of sulphur from the soil. It also kills soil-borne deleterious pests and pathogens.

Burning also results in the loss of important nutrients present in the crop stubble. About 25 % of nitrogen and phosphorus, 50 % of sulphur and 75 % of potassium uptake by cereal crops are retained in crop residues, making them viable nutrient sources (Gadde et al. 2009). According to Singh et al. (2008), nutrient loss due to burning of rice residues in Punjab in 2001–2002 was 2,400 kg of carbon, 35 kg of nitrogen, 3.2 kg of phosphorus, 21 kg of potassium and 2.7 kg of sulphur in 1 ha. While loss of carbon and nitrogen was almost total, the loss of phosphorus, potassium and sulphur was partial (around 20–60 %) as described in Table 2.13.

2.4.1 International Experience

In the United Kingdom a ban on the burning of crop stubble resulted in a decline in the emission of ammonia from 20 Gg nitrogen per year in 1981 to 3.3 Gg nitrogen per year in 1991. According to Gupta and Sahai (2005), 40–80 % of the wheat crop residue nitrogen is lost as ammonia when it is burnt. Similarly, Samra et al. (2003) observed in the case of New Zealand, a ton of wheat residue burnt releases 2.4 kg of nitrogen. Likewise, sulphur (S) losses from the burning of high sulphur and low sulphur rice crop residues in Australia were 60 and 40 % of sulphur content, respectively. Thus, burning may lead to considerable nutrient loss also. According to a study by Heard and Hay (2006), on an average, 98 to 100 % of the nitrogen, 24 % of the phosphorus and 35 % of the potassium and 75 % of the sulphur was lost through burning in the province of Manitoba in Western Canada.

2.5 Health Impacts of Pollution Due to Residue Burning

According to Gadde et al. (2009), open burning of crop stubble results in the emissions of harmful chemicals like polychlorinated dibenzo-p-dioxins, polycyclic aromatic hydrocarbons (PAH's) and polychlorinated dibenzofurans (PCDFs) referred to as dioxins. These air pollutants have toxicological properties and are potential carcinogens. Furthermore, the release of carbon dioxide in the atmosphere due to crop stubble burning results in the depletion of the oxygen layer in the natural environment causing green house effect. Burning of crop waste also has adverse implications on the health of milk producing animals. Air pollution can result in the death of animals, as the high levels of CO_2 and CO in the blood can convert normal haemoglobin into deadly hemoglobin. There can also be a potential decrease in the yield of the milk producing animals.

Burning of crop stubble has severe adverse impacts especially for those people suffering from respiratory disease, cardiovascular disease. Pregnant women and small children are also likely to suffer from the smoke produced due to stubble burning. Inhaling of fine particulate matter of less than $PM_{2.5}$ μg triggers asthma and can even aggravate symptoms of bronchial attack. According to Singh et al. (2008), more than 60 % of the population in Punjab live in the rice growing areas and is exposed to air pollution due to burning of rice stubbles. As per the same study, medical records of the civil hospital of Jira, in the rice-wheat belt showed a 10 % increase in the number of patients within 20–25 days of the burning period every season.

2.6 Management of Crop Stubble

2.6.1 In Situ Incorporation

Though the crop stubble has various alternate uses but the area which is harvested by using combine harvester is left behind with scattered residues which farmers find difficult to remove from the fields. After combine harvesting farmers are left with only two alternatives, either in-situ incorporation of the remains of crop stubble or open burning in the field. Farmers don't prefer in-situ incorporation as the stubble takes time to decompose in the soil that may adversely affect the wheat productivity because of time loss in sowing. As per the Department of Agriculture of the Punjab government, less than 1 % of the farmers incorporate crop stubble because of more tillage operations required in the case of incorporation than of post burning.

As per Singh et al. (1996), if the rice residue is incorporated immediately before sowing the wheat crop, then the crop yield is significantly lowered because of immobilization of inorganic nitrogen and its adverse effect due to nitrogen deficiency. However, in few studies it was found that wheat yield lowered in the first 1–3 years when the rice stubble was incorporated in the soil 30 days prior to sowing of wheat crop, mainly because of the immobilization of soil nitrogen in presence of crop residues with wide C/N ratio. However, in later years rice stubble incorporation did not affect wheat crop yield.

According to another study by Sidhu and Beri (2005), the best alternative available to burning of rice residue is in-situ incorporation. The results of a

6-year study period showed that if the rice residue is incorporated in the soil 10, 20 or 40 days before sowing the wheat crop, then the productivity of the subsequent wheat and rice crops is not adversely affected. Paddy straw incorporated in wheat did not show a residual effect on the succeeding rice crop. Several reports show similar rice and wheat yields under different residual management practices such as burning, removal, or incorporation (Walia et al. 1995; Singh et al. 1996, Singh and Singh 2001). Singh et al. (1996) reported that the incorporation of paddy straw 3 weeks before sowing significantly increased wheat yield on clay loam soil but not on sandy loam soil. Studies conducted by Sharma et al. (1985, 1987) showed no adverse effect of straw incorporation on the grain yield of wheat and the following rice.

As per a study by Singh et al. (1996), the incorporation of rice residue 3 weeks before sowing the wheat crop actually increased the wheat yield only on clay loam soil and not on sandy loam soil. This study further shows that incorporation of crop residues, increased organic carbon by 14–29 %.

According to Verma and Bhagat (1992), incorporation of rice residue 30 days before sowing of wheat crop resulted in lower wheat yields as compared to the wheat yields when the rice residue is burnt or removed from the fields. Furthermore the incorporation of rice stubble in the soil has favorable impact on the soil's physical, chemical and biological properties such as pH, organic carbon, water holding capacity and bulk density of the soil.

The study conducted by Sidhu and Beri (2005) in Ludhiana over an 11-year period to measure the impact of different residue management practices on soil fertility of a sandy loamy soil is revealed in Table 2.14.

As per Mandal et al. (2004), the effect of different crop residue management practices on the physiochemical properties of the soil for 7 years are given in Table 2.16. Paddy straw was incorporated in the soil. From Table 2.15 it is quite clear that, in-situ incorporation of the rice residue is the best crop residue management practice followed by removal of rice residue from the fields and burning for retaining the nutrients in the soil.

2.6.2 Alternative Uses of Crop Stubble

The crop stubble produced during the harvesting of rice and wheat crops can be used for various alternative uses if it is not burnt. These include use of crop stubble as fodder for animals, use of crop stubble for the generation of electricity, use as input in the paper/pulp industry etc. The use of rice residue as fodder for animals is relatively low in Punjab as compared to the wheat stubble. This is because the rice residue is high in silica content which in turn is not good for animal health. However, very often the crop stubble is treated with urea before it is fed to the animals. As per Badve (1991), treating crop residues with 4 % urea and 45–50 % moisture improves the nutritive value by increasing digestibility, palatability and crude protein content.

Table 2.14 Effect of crop residue management on organic C and total N content of soil under the rice-wheat cropping system

References	Types of crop residues	Duration of study (years)	Residue management	Organic C (%)	Total N (%)
Beri et al. (1995)	Rice straw in wheat	10	Removal	0.38	0.051
			Burned	0.43	0.055
			Incorporated	0.47	0.056
Sharma et al. (1987)	Rice straw in wheat and wheat straw in rice	6	Removed	1.15	0.144
			Incorporated	1.31	0.159
Singh et al. (2004)	Wheat straw green manure, and wheat straw + green manure (GM) in rice	6	Removed	0.38	–
			Incorporated	0.49	–
			GM	0.41	–
			Straw	0.47	–

Table 2.15 Impact of different residue management practices in Ludhiana (Punjab)

Soil property	Crop residue management		
	Burned	Removed	Incorporated
Total P (mg kg^{-1})	390	420	612
Total K (g kg^{-1})	17.1	15.4	18.1
Olsen P (mg kg^{-1})	14.4	17.2	20.5
Available K (mg kg^{-1})	58	45	52
Available S (mg kg^{-1})	34	55	61

Source Sidhu and Beri (2005)

Table 2.16 The effect of different crop residue management practices on the soil

Physiochemical properties of the soil	Residues		
	Incorporated	Removed	Burnt
pH	7.7	7.6	7.6
EC (dSm^{-1})	0.18	0.13	0.13
Organic C (%)	0.75	0.59	0.69
Available N (kg ha^{-1})	154	139	143
Available P (kg ha^{-1})	45	38	32
Available K (kg ha^{-1})	85	56	77
Total N (kg ha^{-1})	2,501	2,002	1,725
Total P (kg ha^{-1})	1,346	924	858
Total K (kg ha^{-1})	40,480	34,540	38,280

Source Mandal et al. (2004)

According to Venkataraman et al. (2006), the use of crop waste as fodder was seen high in states where the crop waste generation was high. Also the crop waste is not transported over long distances because of its low bulk density and high transportation costs. The use of crop waste for domestic cooking ranged from 36 to 67 Tg^{-1} with a 95 % confidence interval uncertainty of 86 % (at 95 % confidence interval). The use of crop waste as thatching material was only minor at 2 % of the generated paddy straw.

2.6.3 Cost of Alternate Uses

The use of crop stubble as fodder for animals or for the generation of electricity requires various on-farm and off-farm operations, including collection, packing, handling, transportation, storage and pre feeding processing. For collection of straw after combining, imported conventional field bailers are available.

According to Owen and Jayasuriya (1989) the bulky nature of the straw makes them expensive to transport even for short distances. According to a study by Gupta et al. (2004), the bailing cost is around Rs. 800 ha^{-1}. The total cost of operation, including bailing, collection, transportation up to a 5-km distance and stacking is Rs. 1,300 ha^{-1} or Rs. 650 t^{-1} of straw. However, the problem with these bailers is that they recover only 25–30 % of the potential straw yield after combining, depending upon the height of plant harvested by combines. According to a study by Owen and Jayasuriya (1989), the use of crop stubble in bio-thermal plants has not been very successful. This is mainly due to the lack of any technical and economic feasibility studies, lack of assured markets for processed by-products, shortage of funds to undertake research and development, etc.

2.6.4 End Use of Rice Residue in Different Districts of Punjab

In Bhatinda district the paddy straw was totally burnt with no other end use. In Amritsar district the use of rice residue for other uses apart from burning is the maximum with 18.2 % being used for fodder, 19.6 % being sold in the market and 9.4 % given to poor landless families. In the Gurdaspur district, 20.6 % of the rice residue is provided to poor landless families, 12.9 % used as fodder almost the rest of the rice stubble burnt. In Patiala district 11.7 % of the rice stubble is used as fodder for animals and 5.9 % sold in the market and the rest 81.5 % being burnt. In the Ferozepore district, 18.8 % of the rice stubble is provided to poor landless families, 8.8 % is incorporated in the soil and the rest 68.1 % is burnt. It can be observed that except Ferozepore district, the rice stubble is hardly incorporated in the soil in rest of the state.

However for the wheat crop, a significant proportion of the stubble is used as fodder for animals, in 7 districts of Amritsar, Bhatinda, Faridkot, Gurdaspur, Kapoorthala, Ludhiana and Sangrur, the average being 47 %. Only in the Gurdaspur district 2.4 % of the wheat stubble is incorporated in the soil.

From the literature it is clear that farmers seldom incorporate rice and wheat stubble in the soil. Wheat stubble is used as fodder for animals, but the usage of rice stubble as fodder for animals is not much. The paper mills procured rice-straw at a rate of Rs. 200–300 t^{-1}. The wheat straw was generally sold after making chaff. The price of chaff varied between Rs. 2,500 and 3,700 ha^{-1}.

According to Singh and Singh (2001), incorporation of cereal crop residues immediately before sowing/transplanting into wheat/rice significantly lowers crop yield because of immobilization of inorganic N and its adverse effect due to N deficiency. However, in few studies, wheat yields were lower during the first one to three years of paddy straw incorporation 30 days prior to wheat planting, but in later years straw incorporation did not affect wheat yields adversely.

The incorporation of the straw in the soil has a favorable effect on the soil's physical, chemical and biological properties such as Ph, organic carbon, water holding capacity and bulk density of the soil. On a long-term basis it has been seen that it increases the availability of zinc, copper, iron and manganese content in the soil and it also prevents the leaching of nitrates. By increasing organic carbon it increases bacteria and fungi in the soil. In a rice-wheat rotation, Beri et al. (1992), Sidhu et al. (1995) observed that soil treated with crop residues held 5–10 times more aerobic bacteria and 1.5–11 times more fungi than soil from which residues were either burnt or removed. Due to increase in microbial population, the activity of soil enzymes responsible for conversion of unavailable to available form of nutrients also increases. Mulching with paddy straw has been shown to have a favorable effect on the yield of maize, soybean and sugarcane crops. It also results in substantial savings in irrigation and fertilizers. It is reported to add 36 kg/ha of nitrogen and 4.8 kg/ha of phosphorous leading to savings of 15–20 % of total fertilizer use.

In cognizance of this fact, Department of Farm Power and Machinery, Punjab Agricultural University has developed Happy Seeder machine to solve the problem of straw management in collaboration with CSIRO Land and Water Resources, Australia, under financial assistance from Australian Centre for International Agricultural Research (ACIAR). The machine is compact and lightweight and is tractor mounted. It consists of two separate units, a straw management unit and a sowing unit. The Happy Seeder can handle the paddy straw and do the sowing job without any tillage. It consists of straw cutting and chopping unit and a sowing drill combined in one machine. It sows the seed of next crop in one operational pass of the field, while retaining the rice residue as surface mulch.

Though there are some apprehensions such as increased chances of rodents, etc., the many advantages of adopting the technology are as under:

1. Allows sowing of wheat even when stubble is standing in the field. This is finally incorporated into the soil.
2. Mulching effect of straw causes weed suppression.

3. Possibility of saving first irrigation by sowing wheat in residual moisture.
4. Leads to conservation of water due to moisture retention. There is no loss of nutrients.

This environment friendly technology will prove a boon to the farmer community and the state can help them in making provision of this tool for improving soil health and environment for sustainable agriculture. The Happy Seeder machine has low adoption because of its high price and less popularity among the farmers. The state although is providing subsidy on Happy Seeder but it needs to make the farmers educated on the various benefits of Happy Seeder machine. The state needs to undertake demonstration of this technology to make the farmers understanding this technology appropriately. There is also need to encourage farmers adopting Happy Seeder by developing cooperatives or farmers groups and provide the facility to the small and marginal farmers through custom hiring basis.

2.7 Summary of the Chapter

The chapter puts forth literature on various aspects of residue generated on the field, chemical composition of the residue, volume of pollution caused by residue burning, adverse impact of burning on human and animal health and various ways of crop stubble management. Burning of farm waste causes severe pollution of land and water on local as well as regional scale. It is estimated that burning of paddy straw results in nutrient losses viz., 3.85 million tonnes of organic carbon, 59,000 t of nitrogen, 20,000 t of phosphorus and 34,000 t of potassium. This also adversely affects the nutrient budget in the soil. Straw carbon, nitrogen and sulphur are completely burnt and lost to the atmosphere in the process of burning. It results in the emission of smoke which if added to the gases present in the air like methane, nitrogen oxide and ammonia, can cause severe atmospheric pollution. These gaseous emissions can result in health risk, aggravating asthma, chronic bronchitis and decreased lung function. Burning of crop residue also contributes indirectly to the increased ozone pollution. It has adverse consequences on the quality of soil. When the crop residue is burnt the existing minerals present in the soil get destroyed which adversely hampers the cultivation of the next crop.

Crop residue burning is one among the many sources of air pollution. The on-field impact of burning includes removal of a large portion of the organic material, denying the soil an opportunity to enhance its organic matter and incorporate important chemicals such as nitrogen and phosphorus, as well as, loss of useful micro flora and fauna. The off-field impacts are related to human health due to general air quality degradation resulting in aggravation of respiratory (like cough, asthma, bronchitis), eye and skin diseases. Fine particles also can aggravate chronic heart and lung diseases and have been linked to premature deaths in people already suffering from these diseases. The black soot generated during burning also results in poor visibility which could lead to increased road side incidences of accident. It is thus essential to mitigate impacts due to the burning of agricultural waste in the open fields and its consequent effects on soil, ambient air and living organisms.

References

Agarwal, S., Trivedi, R. C., & Sengupta, B. (2006). Air pollution due to burning of residues. In *Proceeding of the Workshop on Air Pollution Problems Due to Burning of Agricultural Residues, held at PAU, Ludhiana Organized by the Indian Association for Air Pollution Control in collaboration with the Punjab State Pollution Control Board*. New Delhi: Patiala and the Central Pollution Control.

Agarwal, S., Trivedi, R. C., & Sengupta, B. (2008). Air pollution due to burning of agricultural residue. *Indian Journal of Air Pollution Control, 8*(1), 51–59.

AIT-EEC. (1983). *Evaluation and selection of Ligno-Cellulose wastes which can be converted into substitute fuels*. Project report submitted to EEC, Belgium.

Badarinath, K. V. S., & Chand Kiran, T. R. (2006). Agriculture crop residue burning in the indo-Gangetic Plains—A study using IRSP6 WiFS satellite data. *Current Science, 91*(8), 1085–1089.

Badarinath, K. V. S., Kumar Kharol, S., & Sharma Anu, R. (2008). Long range transport of aerosols from agriculture crop residue burning in indo-Gangetic Plains—A study using LIDAR, ground measurements and satellite data. *Journal of Atmospheric and Solar-Terrestrial Physics, 59*(3), 219–236.

Badve, V. C. (1991). Feeding systems and problems in the Indo-Ganges plain: Case study. In A. Speedy & R. Sansoucy (Eds.), Feeding dairy cows in the tropics. *Proceedings of the FAO Expert Consultation Held in Bangkok, Thailand*, July 7–11, 1989.

Beri, V., Sidhu, B. S., Bahl, G. S., & Bhat, A. K. (1995). Nitrogen and phosphorus transformation as affected by crop residue management practices and their influence on crop yield. *Soil Use and Management, 11*, 51–54.

Beri, V., Sidhu, B. S., Bhat, G. S., & Singh, B. P. (1992). Nutrient balance and soil properties as affected by management of crop residues. In M. S. Bajwa, et al. (Eds.), Nutrient management for sustained productivity (vol. II, pp. 133–135). *Proceedings of International Symposium*. Ludhiana, India: Department of Soil, Punjab Agricultural University.

Bhattacharya, S. C., & Shrestha, R. M. (1990). *Bio-coal technology and economics*. Bangkok, Thailand: RERIC.

Bhattacharya, S. C., Pham, H. L., Shrestha, R. M., & Vu, Q. V. (1993). CO_2 emissions due to fossil and traditional fuels, residues and wastes in Asia. In *AIT Workshop on Global Warming, Issues in Asia*, September 8–10, 1993. Bangkok, Thailand: AIT.

Butchaiah G., Christoph M., & Reiner W. (2009). Rice straw as a renewable energy source in India, Thailand and the Philippines' overall potential and limitation for energy contribution and greenhouse gas mitigation. *Biomass and Bioenergy, 33*, 1532–1546.

Canadian Lung Association. (2007). *Pollution and air quality*. http://www.lung.ca/protect-protegez/pollution-pollution_e.php.

Dobermann, A., & Fairhurst, T. H. (2002). Paddy straw management. *Better Crops International, 16*, 7–11.

Gadde, B., Bonnet, S., Menke, C., & Garivait, S. (2009). Air pollutant emissions from rice straw open field burning in India, Thailand and the Philippines. *Environmental Pollution, 157*, 1554–1558.

Garg, S. C. (2008). Traces gases emission from field burning of crop residues. *Indian Journal of Air Pollution Control, viii*(1), 76–86.

Government of Punjab. (2007). *State of environment*. Chandigarh: Punjab State Council of Science and Technology.

Gupta, P. K., & Sahai, S. (2005). Residues open burning in rice-wheat cropping system in India: An agenda for conservation of environment and agricultural conservation. In I. P. Abrol, R. K. Gupta & R. K. Malik (Eds.), *Conservation Agriculture—Status and Prospects* (pp. 50–54). New Delhi: Centre for Advancement of Sustainable Agriculture, National Agriculture Science Centre.

Gupta P. K., Sahai, S., Singh, N., Dixit, C. K., Singh, D. P., Sharma, C. (2004). Residue burning in rice-wheat cropping system: Causes and implications. *Current Science, 87*(12), 1713–1715.

Heard, J., & Hay, D. (2006). Typical nutrient content, uptake pattern and Carbon: Nitrogen ratios of prairie crops. In *Proceedings of Manitoba Agronomists Conference*.

Koopman, A., & Koppejan, J. (1997). *Agricultural and Forest Residues Generation, Utilization and Availability*. Paper presented at the regional consultation on modern applications of biomass energy, January 6–10, 1997. Kuala Lumpur, Malaysia.

Lal, M. M. (2006). An overview to agricultural burning. In *Proceeding of the Workshop on Air Pollution Problems Due to Burning of Agricultural Residues*, held at PAU, Ludhiana organized by the Indian Association for Air Pollution Control in collaboration with the Punjab State Pollution Control Board, Patiala and the Central Pollution Control Board, New Delhi.

Mandal, K. G., Misra, A. K., Hati, K. M., Bandyopadhyay, K. K., Ghosh, P. K., & Mohanty, M. (2004). Rice residue management options and effects on soil properties and crop productivity. *Food, Agriculture and Environment, 2*, 224–231.

Njie, D. (2006). Energy generation from rice residues: A review of technological options, opportunities and challenges. Review Articles, FAO, Rome Italy. ftp://ftp.fao.org/docrep/fao/010/a1410t/a1410t02.pdf.

Owen, E., & Jayasuriya, M. C. N. (1989). Use of crop residues as animal feeds in developing countries. *Research and Development in Agriculture, 6*(3), 129–138.

Punia, M., Nautiyal, V. P., & Kant, Y. (2008). *Identifying biomass burned patches of agriculture residue using satellite remote sensing data*. 4 Kalidas Road, New Delhi, Dehradun, India: Centre for the Study of Regional Development, School of Social Sciences, Jawaharlal Nehru University, India and Indian Institute of Remote Sensing.

Punjab Pollution Control Board. (2007). *Air pollution due to burning of crop residue in agriculture fields of Punjab*. Assigned and Sponsored by PPCB, Patiala and CPCB, New Delhi. www.envirotechindia.com.

Samra, J. S., Singh, B., Kumar, K. (2003). Managing crop residue in the rice wheat system of the indo-gangetic plain [as a special publication]. *Improving the productivity and sustainability of rice-wheat systems: Issues and impacts*, 173–195 (65, Wisconsin, USA).

Sarkar, A., Yadav, R. L., Gangwar, B., & Bhatia, P. C. (1999). *Crop residues in India*. Modipuram: Project Directorate for Cropping System Research.

Sharma, H. L., Modgal, S. C., & Singh, M. P. (1985). Effect of applied organic manure, crop residues and nitrogen in rice-wheat cropping system in north-western Himalayas. *Himachal Journal of Agriculture Research, 11*, 63–68.

Sharma, H. L., Singh, C. M., & Modgal, S. C. (1987). Use of organics in rice-wheat crop sequence. *Indian Journal of Agricultural Science, 57*, 163–168.

Sidhu, B. S., & Beri, V. (2005). Experience with managing rice residues in intensive rice-wheat cropping system in Punjab. In I. P. Abrol, R. K. Gupta & R. K. Malik (Eds.), *Conservation agriculture: Status and prospects* (pp. 55–63). New Delhi: Centre for Advancement of Sustainable Agriculture, National Agriculture Science Centre.

Sidhu, H. S., Singh, M., Humphreys, E., Singh, Y., Singh, B., Dhillon, S. S., Blackwell J., Bector V., Singh M., & Singh S. (2007). The Happy Seeder enables direct drilling of wheat into rice stubble. *Australian Journal of Experimental Agriculture, 47*, 844–854.

Sidhu, B. S., Rupela, O. P., Beri, V., & Joshi, P. K. (1998). Sustainability implications of burning rice and wheat straw in Punjab. *Economic and Political Weekly, 33*(39), A163–A168.

Sidhu, B. S., Beri, V., & Gosal, S. K. (1995). Soil microbial health as affected by crop residue management. In *Proceedings of National Symposium on Developments in Soil Science* (pp. 45–46), Ludhiana, India, November 2–5, 1995. New Delhi, India: Indian Society of Soil Science.

Singh, K., & Rangnekar, D. V. (1986). Fibrous crop residues as animal feed in India. In M. N. M. Ibrahim & J. B. Schiere (Eds.), *Rice straw and related feeds in ruminants' ration* (pp. 111–116). Kandy, Sri Lanka: Straw Utilization Project.

Singh, R. P., Dhaliwal, H. S., Sidhu, H. S., Manpreet-Singh, Y. S., & Blackwell, J. (2008). Economic assessment of the Happy Seeder for rice-wheat systems in Punjab, India. *Conference Paper, AARES 52nd Annual conference, Canberra.* Australia: ACT.

Singh, Y., Singh, D., & Tripathi, R. P. (1996). *Crop Residue Management in Rice-Wheat Cropping System.* Abstracts of poster sessions. 2nd International Crop Science Congress (p. 43). New Delhi, India: National Academy of Agricultural Sciences.

Singh, Y., & Singh, B. (2001). Efficient management of primary nutrients in the rice-wheat system. In: P. K. Katoke (Ed.), *Rice-wheat cropping system of South Asia: Efficient production management.* Binghamton: Food Products Press.

Singh, Y., Singh, B., Ladha, J. K., Khind, C. S., Khera, T. S., & Bueno, C. S. (2004). Management effects on residue decomposition, crop production and soil fertility in a rice-wheat rotation in India. *Soil Science Society of America Journal, 68,* 320–326.

Smill, V. (1999). Nitrogen in crop production: An account of global flows. *Global Biogeochemical Cycles, 13,* 647–662.

Streets, D. G., Yarber, K. F., Woo, J. H., & Carmichael, G. R. (2003). Biomass burning in Asia: Annual and seasonal estimates and atmospheric Emissions. *Submitted to Global Biogeochemical Cycles, 17*(4), 10.1–10.20.

Strehler, A., & Stutzle, W. (1987). Biomass residues. In D. O. Hall & R. P. Overend (Eds.), *Biomass regenerable energy.* New Jersey: Wiley.

Venkataraman, C., Habib, G., Kadamba, D., Shrivastava, M., Leon, J. F., Crouzille, B., Boucher O., & Streets D. G. (2006). Emissions from open biomass burning in India: Integrating the inventory approach with high-resolution moderate resolution imaging spectro-radiometer (MODIS) active-fire and land cover data. *Global Biogeochemical Cycles, 20*(2), 1–12.

Verma, T. S., & Bhagat, R. M. (1992). Impact of rice straw management practices in yields, nitrogen uptake and soil properties in a wheat-rice rotation in northern India. *Fertilizer Research, 33,* 97–106.

Vimal, O. P. (1979). Residue utilization, management of agricultural and agro-industrial residues of selected tropical crops (Indian experience). In *Proceedings of UNEP/ESCAP/FAO,* Workshop on Agricultural and Agro-industrial Residue Utilization in Asia and Pacific Region.

Walia, S. S., Brar, S. S., & Kler, D. S. (1995). Effect of management of crop residues on soil properties in rice-wheat cropping system. *Environmental Ecology, 13,* 503–507.

Chapter 3
Valuation of the Health Effects

Abstract This chapter measures the value of health effects of air pollution for the Indian rural Punjab, where air pollution problem occurs from crop residue burning. Consumer choice model is used to get the monetary estimates of reduced air pollution level to the safe level. The chapter uses data of 625 individuals collected from a household level survey conducted in three villages in Indian Punjab for 150 households. To obtain the monetary values, Tobit and Poisson models are used to estimate mitigation expenditure and workdays lost equations, respectively. Total annual welfare loss in terms of health damages due to air pollution caused by the burning of rice straw in rural Punjab amounts to Rs 76 millions. If one also accounts for expenses on averting activities, productivity loss due to illness, monetary value of discomfort and utility and additional fertilizer, pesticides and irrigation, the losses would be much higher.

Keywords Air pollution · Residue burning · Mitigation expenditure · Workdays lost · Rural Punjab

3.1 Introduction

Epidemiological studies show that the contamination of air quality increases adverse health impacts (Ostro et al. 1995). Air pollution contributes to the respiratory diseases like eye irritation, bronchitis, emphysema, asthma etc., which not only increases individuals' diseases mitigation expense but also affect their productivity at work. Most of the studies valuing health impacts of air pollution remain confined to urban areas as air pollution is considered mainly the problem of urban areas in developing countries. Though health consequences from burning of agricultural residue are not fully understood, relative short exposure may be more of a nuisance rather than a real health hazard, many of the components of agricultural smoke cause health problem under certain conditions (Long et al. 1998).

© The Author(s) 2015
P. Kumar et al., *Socioeconomic and Environmental Implications of Agricultural Residue Burning*, SpringerBriefs in Environmental Science,
DOI 10.1007/978-81-322-2014-5_3

This chapter attempts to estimate the value of health effects of air pollution for the rural Punjab, where air pollution problems happen due to crop residue burning.

The rice and wheat system (RWS) is one of the widely practiced cropping systems in India. About 90–95 % of the rice area is used under intensive rice wheat system in Punjab. Widespread adoption of green revolution technologies and high yielding variety of seeds increased both, crop as well as crop residue. In the last few decades intensive mechanization of agriculture has been occurring and combine harvesting is one such input, particularly in the RWS. Note that in the RWS a short period of time is available between rice harvesting and wheat plantation and any delay in planting adversely affects the wheat crop. This coupled with combine harvesting compels the farmers to burn the residue to get rid of it. It is estimated that 22,289 Gg of paddy straw surplus is produced in India each year out of which 13,915 Gg is estimated to be burnt in the field. The two states namely Punjab and Haryana alone contribute 48 % of the total and are subject to open field burning (Gadde et al. 2009). Burning of straw emits emission of trace gases like CO_2, CH_4, CO, N_2O, NO_X, SO_2 and large amount of particulates which cause adverse impacts on human health. It is estimated that India annually emits 144,719 Mg of total particulate matter from open field burning of paddy straw (Gadde et al. 2009).

There are many studies in developed countries that estimate the value of adverse health effects of air pollution (Gerking and Stanley 1986; Dockery et al. 1993; Schwartz 1993; Pope et al. 1995 etc.). Similar evidences are available from India and other developing countries (e.g., Cropper et al. 1997; Kumar and Rao 2001; Murty et al. 2003; Gupta 2008; Chesnut et al. 1997; Alberini and Krupnick 2000). These studies used either household health production model or damage function or cost of illness approaches to estimate the monetary value of health damage caused due to ambient air pollution. Note that these studies are restricted to measure the monetary value of reducing urban air pollution to the safe level since air pollution has been considered mainly the problem of urban areas.

Cropper et al. (1997) using dose-response model find that a 100-μg/m^3 increase in total suspended particulate matter (TSPM) leads to 2.3 % increase in trauma deaths in Delhi. Kumar and Rao (2001) estimated the household health production function using data of working individuals of the residential complex of Panipat Thermal Power Station in Haryana, India and find that individual willingness to pay varies between Rs. 12 and 53 per month for improving the air quality to WHO standards. Using a similar model, Murty et al. (2003) observed that a representative household gains about Rs. 2,086 and 950 per annum due to reduced morbidity from reduction of air pollution to the safe level in Delhi and Kolkata, respectively. Similarly, Gupta (2008) estimates aggregate benefits of the magnitude of Rs. 225 million per year reducing air pollution to the safe level for the city of Kanpur, India. It is however, to be noted that most of these studies remained confined to only Indian urban areas. In the present study, we use a similar consumer choice model to get the monetary estimates of reduced air pollution level to the safe level for the rural Punjab.

We use data of 625 individuals collected from a household level survey conducted in three villages, namely Dhanouri, Ajnoda Kalan and Simro of Patiala district of Punjab for 150 households. To get the monetary values we estimate two equations: one with mitigation expenditure and the other with workdays lost as dependent variables. Tobit and Poisson models are found to be suitable for estimating mitigation expenditure and workdays lost equations, respectively. We find that total annual welfare loss in terms of health damages due to air pollution caused by the burning of paddy straw in rural Punjab amounts to Rs. 76 millions.

The chapter is organized as follows: Sect. 3.2 presents the ambient air quality levels in the study villages during the period when harvesting of rice takes place. Section 3.3 gives details of the design of the household survey and analyzes the households' behavior. Section 3.4 presents some details of the agricultural output and productivity among the selected households and some health indicators. Section 3.5 describes the theoretical model and estimation strategy of mitigation expenditure and workdays lost function. The results are discussed in Sect. 3.6 while last Sect. 3.7 concludes the chapter.

3.2 Ambient Air Quality Level in Study Area

In Punjab, it is common practice to openly burn agricultural residues in fields after harvesting crops by mechanical harvests. Central and State Pollution Control Boards have been monitoring the ambient air quality for certain Indian cities for the last two decades. Monitoring of ambient air quality in rural areas is very sporadic and purpose specific. Given the severity of the problem, the Punjab Pollution Control Board (PPCB) conducted air pollution monitoring in three villages of Patiala districts, namely Dhanouri, Simro and Ajnoda Kalan during November 1–3, 2006. Patiala is one of the agriculturally leading districts of Punjab which is rich in crops like rice and wheat. When monitoring was done in these designated villages, it was ensured that some burning of paddy straw was happening in the fields of these villages. Monitoring stations for the sites were planned keeping in view the metrological conditions and environmental settings in terms of habited and non-habited areas and following parameters were monitored: metrological parameters (temperature, humidity, wind-speed and wind direction), particulate matters ($PM_{2.5}$, PM_{10}, TSPM), gaseous pollutants (SO_2, NO_X, NH_3, CO, Ozone, THC, TC and BTX) and heavy metals.[1]

Descriptive statistics of some of the important pollutants and metrological parameters is given in Table 3.1. The table shows that gaseous pollutants such as SO_2 and NO_X were within the safe limits put under National Ambient Air Quality Standards (NAAQS) and particulate matters either were measured in terms of SPM,

---

[1] For details, please see the report prepared by Envirotech Instruments Pvt Ltd (2006) on 'Air Pollution Discharged from the Burning of Crop Residue in Agriculture Fields of Punjab' for Punjab State Pollution Control Board, www.envirotechindia.com.

Table 3.1 Descriptive statistics of emissions and metrological data

	PM_{10} ($\mu g/m^3$)	SO_2 ($\mu g/m^3$)	NO_X ($\mu g/m^3$)	Relative humidity (min) (%)	Wind speed (km/h)	Temperature (maximum) (°C)	Temperature difference (maximum minus minimum) (°C)
Mean	306.66	14.20	56.08	46.13	1.78	29.35	16.87
Standard deviation	16.60	2.96	12.32	0.97	0.40	0.60	1.05
Maximum	325.5	17.4	69	47.4	2.35	30.2	18.2
Minimum	284.75	10.25	39.25	45.1	1.425	28.8	15.6

Source PPCB (2007)

PM_{10} or $PM_{2.5}$ and they cross the limits set by the NAAQS.[2] In the study area all the particulates followed same pattern in whatever terms they are measured. The hourly peak values ranged between 300 and 350 and 24 h average concentration ranged between 200 and 300 $\mu g/m^3$. The ratio between peak and average was found to be about 1.2, indicating almost uniform concentration over the monitoring period. The contribution of the burning to PM_{10} concentration appeared to be around 100–200 $\mu g/m^3$. In all the three monitoring sites, the difference in humidity and temperature levels was negligible and the wind speed was found to be in the range of 0–3.6 km/h. Low wind speed coupled with low wind direction fluctuation implies that the impact of polluting activities remain confined to its close vicinity.

3.3 Household Survey Design and Data

To measure the economic cost of pollution, we needed data on other socio-economic-health indicators in addition to pollutant exited by paddy waste burning in the environment. The health indicators during and after the period of burning, measures adopted by people in the periphery to cope with the situation and other socio, agriculture, income and expenditure parameters were not collected by the PPCB survey. In order to further work on the economic cost, we resurveyed the same villages where PPCB conducted its exercise of measuring air pollutant before and after the burning. This exercise assumes that the findings of the PPCB exercise are still valid and no significant change has occurred neither in the incidences of burning nor in the pollutant emitted into the air by the exercise of burning in those areas.

---

[2] CPCB has defined the NAAQS implying the safe level of pollutants for residential, rural and other areas as follows: SO_2 60 and 80, NO_X 60 and 80 and PM_{10} 60 and 100 $\mu g/m^3$ as annual and 24 h averages, respectively.

Table 3.2 Total sown area and area under rice and wheat in the selected villages in the year 2001

Village	Total sown area (acres)	Area under rice (acres)	Area under wheat (acres)
Ajnouda Kalan	897	365	367
Dhanouri	343	156	130
Simbro	755	342	344

Source Official website, district Patiala

Looking at some of the household and agricultural characteristics, the Nabha Tehsil under Patiala District has a total population of 2,51,326 with 75.33 % of it confined in the rural sector and 24.67 % in the urban sector. Out of operated area 26,395 acres of land is under wheat and 28,359 acres of land is under rice crop in the Nabha Tehsil (Table 3.2). Data on the health status and socio-economic variables of households for this exercise were collected through a household survey conducted for this study for the above mentioned three villages in the month of May 2009 and is based on the recall memory. Selection of households in the respective villages was based on stratified random sampling.

The selection of villages was purposive as has been documented above. After selecting the villages, a list of all households including those who were cultivators, agricultural labourers and those who were working in the other formal or informal sectors like regular government or private services, self-business and pension holders was worked out. Stratification was done for the cultivating households in terms of marginal farmers (≤2.5 acres); small farmers (2.51–5.0 acres), medium farmers (5.01–10.0 acres) and large farmers (above 10.0 acres). The farmers were selected on the basis of stratified random sampling method. From each village, approximately 10 farmers were selected for each category.[3] Thus, total 40 farmers were selected from each village and 120 farmers were selected from all the three villages. In addition to cultivating households, a total number of 10 landless labourers were selected from every village. Thus total 30 numbers of agricultural labourers were selected from all the three villages. Therefore, the aggregate sample consists of a total number of 150 households surveyed for this exercise. The division of selected households from different categories is presented in Table 3.4.

The questionnaire used for the household survey had twelve sections seeking detailed information on various aspects of inputs-outputs used by the farmers for agricultural practices, disposal of crop residues and socio-economic characteristics. Section 1 and 2 of the questionnaire provide information on individual household members' profile in terms of their age, education sex, occupation, marital status etc. Section 3 deals with the information on end use of straw and health effects of the burning. It gives information on the current health status of individuals,

---

[3] This is purposive selection and may bias the results, but no other alternative to get the required information. Anyhow, this being the first systematic study, at least provides some estimates of the direction on impacts on human health.

symptoms of illnesses linked to air pollution exposure, averting and mitigation activities followed by all the members in a household during the designated months when the paddy straw burning was happening. Sub-sections also provide information on whether a particular individual is suffering from any chronic diseases. There was question also on the general awareness of households about the illnesses that occur due to air pollution. Sections 4–7 seek information regarding agriculture productivity, input usage, stubble management etc. Section 8 and 9 provide information on medical expenditure and workdays lost during the designated period of paddy straw burning. The former section provides information on the expenditure on formal medication such as fee paid to a doctor, expenses on the allopathic medicines, cost of hospitalization etc. while the later section provides information on expenditure incurred on informal medicines that Indian household generally take without consulting any medical professional. Last two sections provide information on individual habits and households assets. Information on the habits includes whether an individual is habitual to smoking, alcohol drinking or/ and taking any other toxicants and affect health in general.

3.4 The Survey Results

3.4.1 The Household and Farming Characteristics

The phenomenon of adverse gender ratio whereby male members exceed female members in Punjab is also reflected in our selected sample. Out of the total number of households surveyed in our sample, the ratio of male to female was approximately 55–45 % in all the three villages selected (Table 3.3). Population in working age out of total population of all the surveyed households was 70 %. About 28 % of the surveyed household members in the three villages were illiterate or educated up to primary level only. The highest (64 %) households were educated up to secondary level while only 7 % had higher education above secondary level. 67 % household members were self-employed in farming while 14 % were wage earners and 11 % were involved in formal or informal salaried work.

The average size of holdings was around 7 acres among the selected villages out of which 5.4 acres were irrigated (Table 3.4). On an average, leased-in area was more than leased-out area among the selected households, except in Ajnauda Kalan where leased-out area exceeded leased-in area. The cropping intensity was around two crops in a year among all the selected households. On an average, household assets valued at Rs. 3.5 lakhs that included ownership of tractor, submersible pump set, milch and non milching animals and animal house (Table 3.5). Economics of farming is worked out in Table 3.6.

In addition to rice and wheat grown during *kharif* and *rabi* seasons, some area was also devoted to green fodder crops like *jowar* and *bajra* during the *kharif* season and *barseem* during the *rabi* season. There were some miniscule examples of one or two farmers growing mustard (oilseed), moong (pulse crop), sugarcane and maize.

Table 3.3 Household characteristics (%)

Village name	Family size (No.)	Male in the family	HHs in working age (16–60)	Illiterate	Up to secondary	Above secondary	Self employed in farming	Self employed in non farming	Salaried and pensioners	Wage earners
Dhanouri	5.56	54.68	71.22	26.98	66.55	6.47	67.39	9.78	2.17	20.65
Ajnauda Kalan	5.54	55.23	67.87	29.24	62.82	7.94	67.01	11.34	11.34	10.31
Simro	5.92	53.38	70.95	28.38	63.85	7.77	67.92	1.89	17.92	12.26
Total	5.67	54.41	70.04	28.20	64.39	7.40	67.46	7.46	10.85	14.24

Table 3.4 Farm holding characteristics

Village name	Average operated area (acres)	Leased-in area (acres)	Leased-out area (acres)	Irrigated area (acres)	Cropping intensity	Approximately No. hh in the village	No of selected farmers			
							Marginal	Small	Medium	Large
Dhanori	6.07	4.13	2.00	4.74	1.99	117	10	11	13	6
Ajnauda Kalan	8.43	4.45	5.50	6.12	2.00	380	10	10	10	10
Simro	6.86	3.31	0.00	5.17	2.15	152	10	10	15	5
Total	7.12	4.00	3.40	5.36	2.05	649	30	31	38	21

Table 3.5 Value of assets holding among the farmers (Rs. per household)

	Dhanori	Ajnauda Kalan	Simro	Aggregate
Tractor/trolley	111,700	82,700	122,960	105,787
All other mechanical implements	41,162	46,448	88,658	58,756
Happy seeder	2,000	0	70	690
Thresher	40	1,844	2,858	1,581
Irrigation pump sets submersible	59,500	89,900	40,700	63,367
Irrigation pump set non-submersible	9,660	200	200	3,353
Other mechanized assets	800	3,200	120	1,373
Animal house	27,700	42,300	48,650	39,550
Milching animals	57,400	63,380	56,880	59,220
Non milch animals	14,230	15,320	18,000	15,850
Any other	200	2,060	0	753
Total	324,392	347,352	379,096	350,280

The input-output table is worked out for the major crops of rice, wheat and green fodder. Among these three crops, returns were highest in wheat, followed by rice and green fodder while the latter was mostly produced for domestic animals and the crop was not sold in the market. Irrigation, labour, machine cost and plant protection were the major items of cost of production in rice and wheat while seed was the second significant item in the case of green fodder. The net returns per acre were around 14 thousand for rice and 16.5 thousand for wheat. However, while calculating cost of production, we have not included the rental value of own land, interest paid on fixed capital and depreciation value of implements and therefore value of net returns is exaggerated.

3.4.2 Management of Stubble Among the Selected Farmers

Table 3.7 presents the details of total amount of stubble (by-product) generated on the field, its various uses and various alternatives adopted by the farmers to dispose of the stubble for the two main crops of paddy and wheat. On an average, total amount of stubble generated for paddy and wheat per acre was around 23 and 19 quintals, respectively. Out of this in the case of paddy, more than 85 % was burnt in the open field and less than 10 % was incorporated using rotavator while rest of 8 % was used for other purposes. In the case of wheat, 77 % of the total amount was used as fodder for animals while 9 % was incorporated and around 11 % was burnt. The reason for burning high amount of paddy stubble in the open field was non availability of any machine which can be used to collect the crop remains after the combine harvest.[4] Rotavator was used by around 10 % of the

---

[4] High wage rate in the state could be another important reason for use of machines and burning of straw.

Table 3.6 Input-output table (Rs. per acre)

Village name	Seed	Manure fertilizer and pesticides	Irrigation	Labour cost including family labour	Cost of machinery	Rent for leasing-in	Other cost	Total cost	Value of total output (+byproduct)	Net returns
Rice										
Dhanori	303	1,527	2,138	1,908	1,924	0	63	7,863	22,191	14,327
Ajnauda Kalan	276	1,493	2,138	2,065	2,133	385	56	8,547	22,454	13,907
Simro	237	1,183	2,064	2,386	1,896	0	185	7,951	21,706	13,755
Total	272	1,401	2,113	2,120	1,984	128	102	8,120	22,117	13,997
Wheat										
Dhanori	853	1,364	953	2,235	1,993	0	138	7,534	24,427	16,893
Ajnauda Kalan	912	1,528	955	2,509	2,031	38	65	8,037	24,924	16,887
Simro	876	1,098	953	2,078	1,821	0	184	7,009	23,001	15,992
Total	880	1,330	953	2,274	1,948	13	129	7,527	24,117	16,590
Green fodder										
Dhanori	1,091	215	400	1,538	449	0	0	3,692	6,982	3,289
Ajnauda Kalan	1,098	269	364	1,436	409	0	4	3,580	8,038	4,458
Simro	1,040	200	395	1,156	743	0	0	3,534	7,072	3,538
Total	1,076	228	386	1,377	534	0	1	3,602	7,364	3,762

Note Total cost does not include, rental value of own land, depreciation of capital and implements and interest on fixed capital

Table 3.7   The amount of stubble generated on the field and its alternate uses

Variable name		Dhanori	Ajnauda Kalan	Simro	Aggregate
Paddy					
Qty of total residue generated (quintals per acre)		22.61	23.50	21.54	22.55
Percentage of residue	Burnt out of the total	87.54	89.47	78.80	85.48
	Incorporated using HS	0.00	0.00	0.00	0.00
	Incorporated using rotavator	5.82	5.26	15.31	8.59
	Sold in the market	0.00	0.00	2.51	0.78
	Used as fodder	0.00	0.00	0.00	0.00
	Used in bio-thermal plant	1.75	2.24	2.51	2.16
	Used in pulp industry	0.00	0.56	0.00	0.20
	Used for other purposes	4.89	2.46	0.88	2.79
Wheat					
Qty of total residue generated (quintals per acre)		19.33	18.98	17.73	18.68
Percentage of residue	Burnt out of the total	10.87	8.17	13.12	10.66
	Incorporated using HS	0.00	0.00	0.00	0.00
	Incorporated using rotavator	10.09	10.80	6.35	9.15
	Sold in the market	1.94	3.29	3.53	2.90
	Used as fodder	76.33	76.42	77.01	76.57
	Used in bio-thermal plant	0.00	0.00	0.00	0.00
	Used in pulp industry	0.00	0.00	0.00	0.00
	Used for other purposes	0.78	01.32	0.00	0.71

households to incorporate rice remains after combine harvest but it could be used only after partly or fully burning the field. On the other hand, in the case of wheat, 90 % of households used reaper driven by tractor after combine harvest to collect the wheat remains that were used as fodder for animals. Among the selected households, less than 7 % in wheat and 2 % in paddy harvested manually to collect the stubble from the field. More than 80 % households in wheat used reaper to collect wheat residue to use it as fodder. In the case of paddy, around 8 % households indicated using machine (rotavator) to remove crop remains from the field while around 90 % households fully burnt the field to remove the crop residue (Table 3.8).

Providing a reasoning why the majority of the households were burning rice residue, around 41 % households indicated that there was shortage of time period between harvesting of paddy crop and sowing of wheat crop. A majority, 48 % indicated that burning was more economical and only around 8 % opined that they were indulging in burning because there was shortage of labour force for manual removal of the residue (Table 3.9). Shortage of time in wheat sowing after harvest of paddy is further confirmed from the data shown in Table 3.10. Among the

Table 3.8 Residue removal practices in the field (% of households)

Methods of residue removal		Manually	By machine	By burning
Dhanori	Rice	2.35	4.71	92.94
	Wheat	5.62	91.01	3.37
Ajnauda Kalan	Rice	0.00	7.14	92.86
	Wheat	5.81	82.56	11.63
Simro	Rice	3.75	13.75	82.50
	Wheat	9.41	69.41	21.18
Total	Rice	2.01	8.43	89.56
	Wheat	6.92	81.15	11.92

Table 3.9 Motivation for burning of crop residue (% of households)

Motivation to burn residue	Shortage of time between harvesting and next crop	Shortage of labour for manual removal	No economical use of crop residue	Burning is more economical
Dhanori	37.80	7.32	0.00	54.88
Ajnauda Kalan	38.75	10.00	5.00	46.25
Simro	48.53	5.88	1.47	44.12
Total	41.30	7.83	2.17	48.70

Table 3.10 Average no of days available for the next crop when crop residue is removed by different practices

Village/crop		Average no of days available for next crop when residue is			
		Burnt in the case of paddy/use of reaper in the case of wheat	Incorporated in soil using happy seeders	Incorporated in soils using rotavator/zero drill	Removed from the field by other means
Ajnauda Kalan	Rice	46	–	44	44
	Wheat	13	–	18	21
Dhanori	Rice	45	–	49	48
	Wheat	13	–	10	–
Simro	Rice	60	–	–	48
	Wheat	13	–	14	16
Total	Rice	48	–	46	47
	Wheat	13	–	15	18

households who burnt the field the average time of sowing interval was 13 days while households who incorporated stubble using rotavator/zero drill, the average interval was 15 days. The average interval went up to 18 days among those households who used manual or other methods of stubble removal. In the case of wheat, however, there was sufficient time of 46–48 days for removal of stubble before the transplantation of paddy putting no additional compulsion on the farmers to burn the field for removal of crop residue. According to 82 % of the selected farmers, the easiest and quickest way of paddy stubble removal was burning while 14 % indicated incorporation using rotavator (Table 3.11). Not only the majority of farmers were of the opinion that burning was the quickest way of stubble removal but they were also convinced that this method of stubble management was ensuing them the maximum crop yield (Table 3.12).

A few farmers who incorporated the stubble of paddy, they used either rotavator or zero drill machines. The happy seeder machine (new invention for incorporating rice stubble) was not yet adopted by any selected farmers (Table 3.13). However, although farmers were convinced that burning was not harming the level of crop yield but they pointed out that burning of field added extra cost to the production because of top soil getting affected by the burning. Some of these farmers observed changes in colour of the top soil on the surface of the land after burning. Farmers indicated that in comparison to incorporation, burning required, on an average, 20–50 kg of extra chemical fertilizer that added Rs. 250–300 per acre extra cost of production (Table 3.14). The farmers who burnt the field (fully or partly) to clear the wheat stubble used 169 kg of urea in the next crop of paddy while those who incorporated or adopted other means used 145 and 148 kg of urea, respectively. Similarly, those farmers who burnt paddy field, used added amount of Di-Amonia Phosphate (DAP) to recapture the nutritive lost in the fire in comparison to those who incorporated or removed stubble manually (Table 3.16). Higher expenses are not only in terms of higher fertilizer but also in terms of higher irrigation requirement by those who burn their field to clear the stubble as indicated by Table 3.15. In paddy the total cost of irrigation was Rs. 2,220 for those who burn their filed in comparison to Rs. 2,000 for the farmers who incorporated the wheat stubble. In the case of wheat, per acre irrigation cost was Rs. 941 for those who burned the field

Table 3.11 The easiest and quickest way to get rid of the crop stubble (% of households)

Crop stubble removal method	Burning	Incorporation using happy seeder technology	Incorporation using other methods	Manual way of harvesting and collection	Removal from the field by other means
Dhanori	89.89	0.00	8.99	1.12	0.00
Ajnauda Kalan	79.55	0.00	14.77	3.41	2.27
Simro	78.82	0.00	17.65	3.53	0.00
Total	82.82	0.00	13.74	2.67	0.76

Table 3.12 Households' perception about which method of crop stubble management gives them the maximum crop yield (% of households)

	Burning	Happy seeder technology	Zero drill/rotavator	Traditional way
Dhanori	83.15	0.00	15.73	1.12
Ajnauda Kalan	71.59	0.00	28.41	0.00
Simro	74.12	0.00	23.53	2.35
Total	76.34	0.00	22.52	1.15

Table 3.13 If crop subtle incorporated in the soil, method used for incorporation (% of households)

Incorporation method	Happy seeder	Zero till drill	Rotavator	Manually
Dhanori	0.00	2.78	88.89	8.33
Ajnauda Kalan	0.00	7.16	92.86	0.00
Simro	0.00	4.00	92.00	4.00
Total	0.00	4.49	91.01	4.49

Table 3.14 Additional fertilize use when crop stubble burning

	Higher amount of fertilizer required for next crop when crop subtle of previous crop burnt			Difference seen on surface of top of soil when residue burnt (% of households)
	Percentage HH	Extra amount of Fertilizer (kg per acre)	Extra amount of Fertilizer (value Rs. per acre)	
Dhanori	14.29	46.67	233.33	14.29
Ajnauda Kalan	10.00	23.00	247.50	17.50
Simro	23.08	53.89	305.56	12.82
Total	15.70	45.11	270.53	14.88

but slightly less, Rs. 907 for those who removed stubble by other means. However, in wheat the cost of irrigation was higher for those who incorporated probably because of additional irrigation requirement for stubble fixation.

3.4.3 The Effect of Crop Stubble Burning on Human Health

As mentioned in the beginning of the chapter, air pollution leads to respiratory diseases like eye irritation, bronchitis, asthma etc., increasing individuals' disease mitigation expenses and also affecting ones' working capacity. In addition, open burning in the field affects life of animals, birds and other insects below and above the earth. Burning at times also causes poor visibility and increases the incidents of road accidents. Our household survey shows that paddy stubble burning among our selected villages leads to air pollution and several other problems. There was no conclusive evidence of smoke caused by stubble burning affecting the health or productivity of the milk producing animals (Table 3.17). On the other hand, a significant numbers of households indicated that smoke caused loss to the vegetation in the field and it also led to accidents taking place on the road during the peak of stubble burning that happens in the months of October and November every year. To our question whether households were aware of harmful effects of residue burning, more than 90 % selected households indicated yes but almost none of them was taking any preventive measures to escape from smoke disease before the beginning of the harvest season.

Table 3.15 The effect of end use of straw on the amount of irrigation used per acre

Village name	End use of previous crop residue	Electric tube well—owned and hired			Diesel tube well		Total charges (Rs.)
		Submersible (No. of hours)	Non submersible (No. of hours)	Charges (Rs.)	No. of hours	Charges (Rs.)	
Paddy							
Dhanori	Residue burnt	15	–	1,000	10	1,500	2,500
	Incorporated	–	–	–	–	–	–
	Other uses	17	20	727	10	1,423	2,150
Ajnauda Kalan	Residue burnt	15	–	600	10	1,500	2,100
	Incorporated	15	–	500	10	1,500	2,000
	Other uses	17	–	781	9	1,367	2,148
Simro	Residue burnt	10	–	833	9	1,350	2,183
	Incorporated	–	–	–	–	–	–
	Other uses	14	–	653	9	1,341	1,994
Total	Residue burnt	12	–	820	9	1,400	2,220
	Incorporated	15	–	500	10	1,500	2,000
	Other uses	16	20	722	9	1,378	2,100

(continued)

Table 3.15 (continued)

Village name	End use of previous crop residue	Electric tube well—owned and hired			Diesel tube well		Total charges (Rs.)
		Submersible (No. of hours)	Non submersible (No. of hours)	Charges (Rs.)	No. of hours	Charges (Rs.)	
Wheat							
Dhanori	Residue burnt	7	10	348	4	563	911
	Incorporated	8	0	500	4	600	1,100
	Other uses	–	–	–	5	700	700
Ajnauda Kalan	Residue burnt	8	–	374	4	611	985
	Incorporated	–	–	–	5	675	675
	Other uses	–	–	–	4	600	600
Simro	Residue burnt	7	–	364	4	561	925
	Incorporated	6	–	313	4	650	963
	Other uses	6	–	200	6	825	1,025
Total	Residue burnt	7	10	362	4	579	941
	Incorporated	7	–	375	4	644	1,019
	Other uses	6	–	200	5	707	907

Table 3.16   The effect of end use of straw on the amount of fertilizer used per acre

Village name	End use of previous crop residue	Manure (Rs.)	Chemical fertilizer (NPK)			
			Urea (kg)	DAP (kg)	MOP (kg)	Value [a]total (Rs.)
Paddy						
Dhanori	Residue burnt	167	167	–	–	1,333
	Incorporated	–	–	–	–	–
	Other uses	150	144	–	–	1,321
Ajnauda Kalan	Residue burnt	350	150	–	–	1,475
	Incorporated	–	145	–	–	1,350
	Other uses	117	139	–	–	1,276
Simro	Residue burnt	100	183	1	–	1,183
	Incorporated	–	–	–	–	
	Other uses	27	161	–	–	1,293
Total	Residue burnt	188	169	–	–	1,313
	Incorporated	–	145	–	–	1,350
	Other uses	98	148	–	–	1,297
Wheat						
Dhanori	Residue burnt	185	131	106	–	1,813
	Incorporated	–	150	100	–	1,690
	Other uses	120	130	100	–	1,760
Ajnauda Kalan	Residue burnt	103	143	94	–	1,759
	Incorporated	250	125	100	–	1,815
	Other uses	–	138	100	–	1,728
Simro	Residue burnt	16	150	91	–	1,629
	Incorporated	100	158	84	–	1,697
	Other uses	–	125	75	–	1,400
Total	Residue burnt	103	141	97	–	1,736
	Incorporated	122	150	89	–	1,722
	Other uses	55	132	95	–	1,683

[a] Total value also includes the expenditure incurred on other chemicals like zink etc., used by the farmers

Irritation in eyes and congestion in the chest were the two major problems faced by the majority of the household members (Table 3.18, Fig. 3.1). Respiratory allergy, asthma and bronchial problems were the other smoke related diseases which affected household members in the selected villages. Almost 50 % of the selected households indicated that their health related problems get aggravated during or shortly after harvest when crop stubble burning is in full swing during the months of October, November and December. In the peak season, affected families had to consult doctor or use some home medicine to get relief from irritation/itching in eyes, breathing problem and similar other smoke related problems. On an average, the affected members suffered at least half a month from such problems and had to spend Rs. 300–500 per household on medicine

Table 3.17 Percentage of household experiencing any problem due to smoke caused by crop stubble burning (mainly rice)

Village/nature of problem	Awareness of harmful effect of residue burning	Family taken preventive measure to escape from smoke disease	Smoke affected productivity of other working members (attendant)	Experienced decline in productivity of milk producing animal	Milk producing animal suffering from sickness	Loss of vegetation due to smoke	Having observed accident happening due to smoke
Dhanori	87.50	2.04	0.00	6.00	0	60.00	8.00
Ajnauda Kalan	92.50	0.00	4.17	0.00	0	59.18	24.00
Simro	95.83	0.00	0.00	2.04	4.08	87.76	89.80
Total	92.19	0.71	1.37	2.68	1.35	68.92	40.27

Table 3.18 Percentage of HH members suffering from the disease due to stubble burning

Problem	Dhanori	Ajnauda Kalan	Simro	Aggregate
Bronchial problems (inflammation of lungs due to infection or other causes)	1.56	2.04	0.00	1.14
Irritation in eyes (eyes feel as being burnt)	75.00	73.47	93.65	81.25
Coughing (congestion in the chest)	34.38	20.41	44.44	34.09
Experience nose/throat irritation due to smoke	4.32	0.00	0.68	1.65
Asthma (shortness of breath, congestion in the chest)	3.13	4.08	23.81	10.80
Emphysema (lung disease due to exposure to smoke, toxic chemicals etc.)	0.00	0.00	0.00	0.00
Respiratory allergies (hay fever caused due to over reaction of the immune system to a stimulus like dust, smoke, air pollutants etc.)	7.81	12.24	15.87	11.93
Other lung and heart disease	0.00	0.00	0.00	0.00
Any other problem	3.13	8.16	1.59	3.98
Health problem gets aggravated during stubble burning	52.00	52.00	42.55	48.98

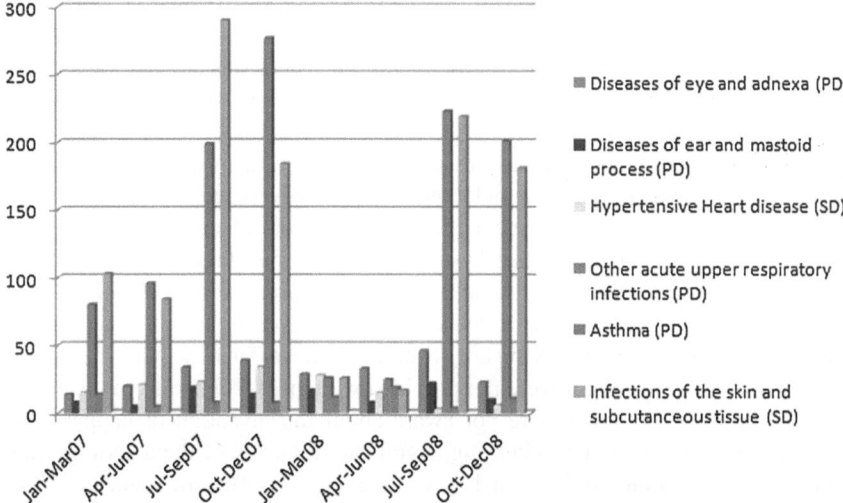

Fig. 3.1 Number of patients treated in the village dispensary Ajnauda Kalan. *Source* Based on primary information collected from village dispensary Ajnauda Kalan by our field survey team

(Table 3.19). In addition there were few examples where a family member had to be hospitalized for three to four days and additional expenditure was incurred. Table 3.20 presents the total medical expenses incurred due to health problem caused by crop stubble burning. Members suffering from smoke related chronic

Table 3.19 Expenditure incurred due to problems faced during the crop stubble burning

Problem		Dhanori	Ajnauda Kalan	Simro	Aggregate
Family members visited local doctor during Oct–Nov, 2008	Average no of members per household	2.93	2.12	2.82	2.63
Prescribed to any medicine during the 2 months of stubble burning (Oct–Nov, 2008)	Average no of members per hh	2.93	2.15	2.82	2.64
	Avg no of days per hh	13.3	13.75	11.43	12.92
	Avg amount spent per hh Rs.	280.33	335.77	504.76	360.26
Any member hospitalized during the 2 months of stubble burning (Oct–Nov, 2008)	Average no of members per hh	0.00	1.00	3.00	2.00
	Avg no of days per hh	0.00	3.00	5.00	4.00
	Avg amount spent per hh Rs.	0	300	1,000	650

and non-chronic diseases observed that their problem becomes acute and the severity increases during the time of crop stubble burning. On an average, households spent around more than a thousand Rupees on the non-chronic respiratory diseases like coughing, difficulty in breathing, irregular heartbeat, itching in eyes decreased lung function etc., during the year 2008–2009.[5] However, out of this total expenditure, around 40–50 % was spent during the months of October and November during the time of crop stubble burning. There was an additional cost in terms of household members remaining absent from work due to illness (Table 3.20).

Some respondents pointed out that Punjab government from time to time advises farmers not to set their field on fire. It is advertised in the local newspapers to make people aware about the adverse effects of crop stubble burning. Some respondents pointed out that District Commissioner directed *gram panchayats* to prevent stubble burning (Table 3.21). The administration even makes such announcements by loud speaker in the villages. However, no documentary or road shows were organized in this regard. Similarly, those farmers who incorporate stubble instead of burning it were not provided with any incentive from the administration. Although farmers were not aware about the invention of happy seeder which can provide alternate to burning, significant majority of them showed interest to buy such machine if the Punjab government gives sufficient financial support for such machine given the price of happy seeder exceeds Rs. 1 lakh in the market. The farmers indicated that if at least half of the price is born by the government for happy seeder they would be interested to buy the happy seeder machine.

---

[5] It should not be considered as the full cost of treatment since many of the patients are treated in public funded institutions or government hospitals where treatment is highly subsidized.

Table 3.20 Medical expenses incurred due to health problem caused by crop stubble burning

Village name	Medical Expenses incurred during the last year (Apr 08 to Mar 09)		Percentage of affected members observing severity of problem increasing at the time of crop stubble burning	Medical Expenses incurred due to acute problem during the crop stubble burning (Oct–Nov 08)				Absence from work for each illness during Oct–Nov 2008	
	Chronic disease (Rs. per affected member)	Non chronic disease (Rs. per affected member)		Doctor/ Hospital/Other charges (Rs. per affected member)	Medicine cost (Rs. per affected member)	Transportation/ freight (Rs. per affected member)	Self medication (Rs. per affected member)	(No of days per affected member)	Money loss (Rs. per affected member)
Dhanori	1,667	767	100.00	71	162	77	80	3	300
Ajnauda Kalan	1,750	1,978	97.30	159	222	80	56	5	600
Simro	2,000	478	95.45	116	200	71	92	10	1,000
Total	1,750	1,145	97.75	119	196	77	82	5	610

Table 3.21 Percentage of households saying yes to the following questions

Problem	Dhanori	Ajnauda Kalan	Simro	Aggregate
Are respondents aware of the Punjab government's policies towards pollution?	12.00	16.00	10.20	12.75
Has the Punjab government taken steps to make people aware of the adverse implications of crop stubble burning in open fields?	6.00	4.00	2.04	4.03
Have any seminars/documentary/road show being organized by the Punjab government to make people aware of the harmful effects of crop stubble burning?	0.00	2.00	0.00	0.67
Has the Punjab government given them enough incentives to stop burning the crop waste?	0.00	0.00	0.00	0.00
Are respondents aware of the happy seeder technology for getting rid of the crop stubble?	0.00	0.00	0.00	0.00
Have respondents been provided subsidy to purchase the happy seeder technology?	0.00	0.00	0.00	0.00
Are respondents willing to buy the happy seeder technology if given the financial support by the Punjab Government	73.17	35.42	41.30	48.89
How much amount of subsidy respondents think the Punjab Government should provide given the market price of happy seeder (% amount of subsidy)	50.00	50.00	50.00	50.00

3.5 Methodology

3.5.1 Theoretical Model

Air quality affects the utility of individuals and an economic value exists. There are several ways to capture this economic value, viz., dose-response, revealed preferences and contingent valuation methods. The dose-response method assumes a relationship between air quality and morbidity (and/or mortality). It puts a price tag on air quality without retrieving people's preferences for the good. But, such type of mechanical relationship of the dose response function does not take into account consumer behavior. The revealed preference methods assume that the consumers are aware of the costs/benefits of air quality and are able to adjust their behavior to reveal their preferences. This necessitates the need to have estimates of willingness to pay (WTP) or willingness to accept (WTA) on the basis of a consumer choice models aimed at measuring the strength of association between health effects and contaminated air quality.

Suppose an individual maximizes his/her utility through expenditure on marketed goods and services, X.[6] The utility depends not only on X but also on the state of health, H of an individual which is affected by the level of air quality (non-marketed good). It is further assumed that the contaminated air quality, P is beyond the control of individuals, but individuals can at least partially reduce its effects through incurring defensive expenditure, D. The utility function is defined as:

$$U(X, D; P) = H(D, P)U(X) \tag{3.1}$$

where $U_X > 0$, $U_{XX} < 0$, $H_D > 0$, $H_P < 0$, $H_{DD} < 0$, $H_{PP} < 0$.

The state of health affects individual's work performance and hence the wage income. Moreover, it is also possible that the contaminated air quality make the individual so sick as to be completely incapacitated. During the time the individual is under this condition, he/she is absent from work and loses the wage income completely. Therefore 'sick time', S can also be assumed to be the function of defensive expenditure and contaminated air quality,

$$S = S(D, P) \tag{3.2}$$

where $S_D < 0$, $S_P > 0$, $S_{DD} > 0$, $S_{PP} > 0$

The Eq. (3.1) is maximized subject to the following constraints:

The time constraint is:

$$W + S = T \tag{3.3}$$

where W is the work time and T is the total time available. The income (resource) constraint is:

$$I + wH(D, P)W \geq mS + D + X \tag{3.4}$$

where, mS is the medical expenses which are assumed proportional to illness, S, I denotes non-wage income and w is referred as wage rate.

The Lagrangian of the problem is:

$$\Pi = H(D, P)U(X) + \lambda[I + wH(D, P)(T - S) - D - X - mS] \tag{3.5}$$

The first-order optimization conditions are:

$$\Pi_X = H(D, P)U_X - \lambda = 0$$
$$\Pi_D = H_D U + \lambda H_D wW - \lambda HwS_D - \lambda - \lambda mS_D = 0 \tag{3.6}$$

Using the envelope theorem, Harrington et al. (1989) obtain the individual willingness to pay (WTP) as:

$$WTP = -\left(\frac{H_D U}{\lambda} + H_D wW\right)\frac{H_P}{H_D} + (HwS_D + mS_D)\frac{S_P}{S_D} \tag{3.7}$$

---

[6] Harrington et al. (1989) take the individual utility as a function of expenditure on marketed goods and services, X and leisure time, L. Since in developing countries especially in rural areas people are living in the conditions of poverty, therefore, we assume that the individual utility is the function of marketed goods and services, X only.

and the marginal loss of social welfare (SW) associated with individual responses
to deterioration in air quality, therefore, is:

$$\frac{\partial SW}{\partial P} = -\frac{U}{\lambda}\frac{dH}{dP} \quad \text{(Direct disutility of illness)}$$

$$-w \times W\frac{dH}{dP} \quad \text{(Lost work productivity)}$$

$$-Hw \times W_P \quad \text{(Value of lost time during illness)} \tag{3.8}$$

$$+m\frac{dS}{dP} \quad \text{(Medical expenses)}$$

$$+D_P \quad \text{(Defensive expenditure)}$$

Equation (3.8) shows that the cost of illness caused by the contaminated air can be
grouped into five categories. The term direct disutility is very subjective and it is
very difficult to find its monetary value. The second term, the lost work productivity measures the value of loss caused by the illness due to lower work productivity. This loss is caused when the sick person is present for work but is not able to
work with his/her full productivity. The third term measures the loss in social welfare due to illness absence of individuals from work. The last two terms measure
the expenses individual have to incur for defensive and mitigating activities due to
contamination of air quality. In rural areas during survey we could not get figures
on the defensive activities of individuals, therefore we measures only two values:
medical expenses and value of lost time during illness. Thus our measure of social
loss due to contaminated air provides the lower bound of the value.

3.5.2 Estimation Strategy

To get the estimates of social welfare loss due to contaminated air in terms of
health damages, we estimate the following two equations consisting of demand
function for medical expenses (mS) and the workdays lost due to illness (S):

$$mS = \alpha_0 + \alpha_1 SPM + \alpha_2 SO_2 + \alpha_3 SMOKING + \alpha_4 DRINKING + \alpha_5 PerCapitaAssets$$
$$+ \alpha_6 SEX + \alpha_7 AGE + \alpha_8 EDUCATION + \alpha_9 OCCUPATION + \varepsilon_1$$

$$\tag{3.9}$$

and

$$S = \alpha_0 + \alpha_1 SPM + \alpha_2 SMOKING + \alpha_3 DRINKING + \alpha_4 PerCapitaAssets + \alpha_5 SEX$$
$$+ \alpha_6 AGE + \alpha_7 EDUCATION + \alpha_8 OCCUPATION + \varepsilon_2$$

$$\tag{3.10}$$

where:

Medical Expenses (mS)	Mitigating activities or medical expenses include expenses incurred as a result of air pollution related diseases. These expenditures include costs of medicine (formal as well informal), doctor's fee, diagnostic tests, hospitalization, and travel to doctor's clinic during the rice harvesting 2 months
Workdays Lost (S)	S represent the number of workdays lost per person during the two rice harvesting months of October and November due to diseases/symptoms associated with air pollution
Particulate matter (PM_{10}) and Sulfur Dioxide (SO_2)	These are the averages of the ambient emission levels observed during the monitoring period measured in $\mu g/m^3$
SMOKING	Measured as dummy variable equal to 1 if the individual is having smoking habit, otherwise 0
DRINKING	Measured as dummy variable equal to 1 if the individual is having alcohol drinking habit, otherwise 0
PerCapitaAssets	Measured in Indian rupees
SEX	Measured as dummy variable equal to 1 for male and 0 for female
AGE	Age of the individual measured in number of years
EDUCATION	Is coded as follows: 1 = Illiterate; 2 = below primary; 3 = Primary; 4 = Middle; 5 = Secondary/Metric; 6 = Technical; 7 = Graduate; 8 = Post graduate and above
OCCUPATION	Measured as dummy variable equal to 1 if the individual is in the occupation of self farming or agricultural labourer, 0 otherwise

Note that the dependent variable in Eq. (3.9) is a censored variable, i.e., the dependent variable is zero for corresponding known values of independent variables for part of the sample. Therefore, we use Tobit model for estimating the demand for mitigating activities:

$$mS_i = \alpha + \beta x_i + u_i \quad \text{if } RHS > 0$$
$$= 0 \quad \text{otherwise}$$

(3.11)

where mS_i refers to the probability of the ith individual incurring positive medical expenditure and x_i denotes a vector of individual characteristics, such as assets, age, sex, education, pollution parameter etc.

In Eq. (3.10) the dependent variable is a count of the total number of workdays lost due to air pollution related illness by an individual during the particular period; therefore, there are zeros for many observations. In this case Poisson regression model is appropriate as it considers the predominance of zeros and the small values and the discrete nature of the dependent variable. The least square

Table 3.22 Variables used in the analysis

Variable	Mean	Standard deviation	Maximum	Minimum	Percent
Formal medical expenses	39.26	165.05	2,700.00	0.00	
Informal medical expenses	19.46	66.62	450.00	0.00	
Workdays lost	0.06	0.72	15.00	0.00	
Age	31.35	18.50	90.00	1.00	
Education	3.14	1.77	8.00	1.00	
Per capita assets	64,469	78,377	539,467	250	
Male					54.41
Occupation (farmers and agricultural laborers)					26.32
Smoking					2.12
Drinking					5.88
Toxicants					3.29

and other linear regression models do not take into account these features. The Poisson regression model can be stated as follows:

$$prob(Y_i = y_i/x_i) = \mu_i^{y_i} e^{-\mu_i}/y_i, y_i = 0, 1, 2, \ldots \tag{3.12}$$

This equation is non-linear in parameters; therefore, for estimation purpose by taking its natural log we convert it into an equation which is linear in parameters. Note that the Poisson regression model is restrictive in many ways. For example, the assumption that the conditional mean and variance of y_i, given x_i are equal, is very strong and fails to account for over dispersion.[7] Table 3.22 gives the descriptive statistics of the variables used in the estimation of the models.

3.6 The Model Results

Tables 3.23 and 3.24 provide the results of parameter estimates of reduced form equations of mitigation expenditure and workdays lost. In the reduced form these equations are expressed as functions of a common set of socio-economic variables and ambient air pollution expressed in terms of particulate matter (PM_{10}) and SO_2 levels.

The parameter estimates of mitigating expenditure equation are given in Table 3.23. We find there is a positive and statistically significant (at 10 % level) association between ambient PM_{10} level and the mitigating expenditure.[8] This implies that individual have to spend higher amount of money to mitigate the adverse health effects when the particulate level is higher in the ambient environment. The relationship between mitigating expenditure and ambient SO_2 level is negative and statistically insignificant, as contrary to expectations. This might be happening as the ambient SO_2 level is within the NAAQS limits in the villages of Punjab.

---

[7] Similar estimation procedure is followed by Gupta (2008).

[8] Farmers take precautionary medical expenses in anticipation of the environmental pollution due to straw burning.

Table 3.23 Tobit equation of total medical expenditure (left censored at 0)

Independent variable	Coefficient
PM$_{10}$ (+)	0.046 (1.72)*
SO$_2$ (+)	−5.16 (−0.52)
SMOKING (+)	395.14 (2.62)***
DRINKING (+)	177.94 (1.71)*
Per capita assets (+)	0.0009 (2.65)***
SEX	−41.76 (−0.57)
AGE (+)	4.13 (1.74)*
EDUCATION (−)	−9.85 (−0.51)
OCCUPATION (+)	92.58 (1.16)
Constant	−678.69 (−2.73)***
Pseudo R^2	0.014
Log likelihood	−1,262.37
Wald Chi^2 (9)	35.74***
Uncensored observations: 141	Left censored observations: 484
Total observations	625

Notes Figures in parentheses are t-values
***Significance at 1 % level; **Significance at 5 % level; *Significance at 10 % level

Table 3.24 Poisson equation of workdays lost

Independent variable	Coefficient
PM$_{10}$ (+)	0.008 (5.59)***
SMOKING (+)	−14.66 (−0.01)
DRINKING (+)	−0.81 (−0.79)
Per capita assets (−)	−0.00001 (−1.78)*
SEX	0.43 (1.07)
AGE	−0.011 (−0.97)
EDUCATION (−)	−0.71 (−5.07)***
OCCUPATION	−0.32 (−0.67)
Constant	−5.02 (−3.98)***
Pseudo R^2	0.023
Log likelihood	−170.93
Wald Chi^2 (8)	97.97
Total observations	625

Notes Figures in parentheses are t-values
***Significance at 1 % level; **Significance at 5 % level; *Significance at 10 % level

As is expected, the coefficient of the variables such as smoking and drinking behaviour of the individual are found to be positive and statistically significant. These personal habits coupled with the ambient air pollution make individual more prone to asthmatic diseases and as a result they are required to spend more on mitigating activities. Similarly we find there is positive and significant relationship between the age of individual and their mitigating expenses implying that the marginal effect of age on mitigating expenses is positive. We also observe that there is positive and statistically significant relationship between mitigating expenses and per capita assets. This might be happening because wealthier individuals do not hesitate to take mitigating activities if they are suspected to some diseases in comparison to people who have lesser assets.

Education raises awareness level of individuals with respect to environmental problems and related health damages and helps in taking informed preventing activities related decisions. The coefficient of education is negative, as expected, though statistically insignificant, depicts that there happens to be a reduction in mitigation expenditure with the increase in education level. Similarly, the individuals who have to work in agriculture fields where burning of agricultural residue take place are thought to be more prone to the adverse effects of pollution in comparison to their counterparts who are in other occupations such as salaried individuals. We use dummy variable equal to one for farmers and agricultural wage earners and zero for the individuals who are in other occupations. We find a positive association between occupation variable and medical expenditure.

Table 3.24 presents parameter estimates of the reduced form equation of workdays lost. As expected, the coefficient of PM_{10} variable is positive and statistically significant at 1 % level implying that the probability of losing workdays increases as the concentration of particulate matters in ambient environment increases. Education increases awareness level and helps in taking preventing action and as a result an individual is expected not to lose workday, therefore, we find that there is negative association between education level of individuals and workdays lost. Similarly, wealthier individuals could spend money on preventing activities and there is negative relationship between per capita assets and workdays lost.

3.6.1 Welfare Loss

The welfare loss in terms of health damage due to increase in the concentration of particulate matters from paddy straw burning in the ambient environment can be estimated in terms of increase in the medical expenditure on mitigating activities and the opportunity cost of workdays lost and are presented in Table 3.25.

3.6.2 Increase in Medical Expenditure

To get the estimates of welfare loss in terms of increased medical expenditure we need to obtain the marginal effects. The marginal effects in the case of Tobit estimation could be computed by taking partial derivatives of mitigating expenditure equation with respect to PM_{10} and multiplying it by the probability

Table 3.25 Welfare loss due to increased air pollution in rural Punjab

	Representative individual (Rs.)	Rural Patiala District (Rs. millions)	Rural Punjab (Rs. millions)
Medical expenditure	2.17	2.35	36.52
Opportunity cost of workdays lost	2.35	2.54	39.57
Total welfare loss	4.52	4.89	76.09

of the dependent variable taking the non-zero values. If the ambient PM_{10} level is reduced from the level observed during the harvesting period of rice in rural Punjab to the safe level (i.e., a reduction of 207 $\mu g/m^3$ since the safe level defined under NAAQS is 100 $\mu g/m^3$ for the 24 h average), the estimated reduction in medical expenditure turns out to be Rs. 2.17 for the months of October and November for a representative person.

Total rural population projected for October 2008 based on Census 2001 is 1,083 thousand and 16,839 thousand for the district of Patiala and the state of Punjab, respectively. Extrapolating this welfare loss for the entire rural population of Patiala and Punjab, it is estimated as Rs. 2.35 million and Rs. 36.52 million, respectively.

3.6.3 Opportunity Cost of Increase in Workdays Lost

To get the marginal effects of reduction in PM_{10} level on workdays lost, we differentiated partially the reduced form equation of workdays lost with respect to PM_{10}. The Poisson estimates show that 1 $\mu g/m^3$ increase in PM_{10} results in a marginal loss of 0.0000946 days for a representative individual in these two harvesting months. If the PM_{10} level is reduced from the current level to the safe levels during rice harvesting period, the estimated gain in workdays is 0.03. In monetary terms, the loss in terms of workdays lost for a representative individual is estimated to be Rs. 2.35 and for rural Patiala district and rural Punjab state it turns out to be Rs. 2.54 million and 39.57 million, respectively assuming a wage rate of Rs. 120 per day.[9]

The total monetary loss (due to lost workdays and increased medical expenditures) caused in terms of health damages due to increase in ambient PM_{10} level beyond the safe level for the rural areas of Patiala district and Punjab state is estimated as, Rs. 4.89 million and Rs. 76.09 million, respectively. These losses should be considered the lower bound of health damages caused by the increased air pollution level in rural Punjab. These estimates could be much higher if expenses on

---

[9] A wage rate fixed for the state of Punjab under National Rural Employment Guarantee Act (NREGA).

averting activities, productivity loss due to illness, monetary value of discomfort and utility could also be considered. There is additional monetary cost of burning to the farmers in terms of additional fertilizer, pesticides and irrigation as was shown by the survey results discussed in section 4. One also has to add into the above cost the losses of soil nutrient, vegetation, bio-diversity and accidents caused because of low visibility.

3.7 Summary of the Chapter

In this chapter an attempt is made to estimate the monetary value of health damage caused by the smoke pollution emitted by the burning of rice and wheat stubble in the open fields in Punjab, India. We use data of 625 individuals collected from a household level survey conducted in three villages, namely Dhanouri, Ajnoda Kalan and Simro of Patiala district of Punjab for 150 households. To get the monetary values we estimated two equations: one with mitigation expenditure and the other with workdays lost as dependent variables. Tobit and Poisson models are used for estimating mitigation expenditure and workdays lost equations, respectively.

On an average, total amount of stubble generated for paddy and wheat per acre was around 23 and 19 quintals, respectively. Out of this in the case of paddy, more than 85 % was burnt in the open field and less than 10 % was incorporated, while rest of 8 % was used for other purposes. In the case of wheat, 77 % of the total amount was used as fodder for animals while 9 % was incorporated and around 11 % was burnt. Although farmers were convinced that burning was not harming the level of crop yield but they pointed out that burning of field added extra cost to the production because of top soil getting affected by the burning. The farmers who burnt the field (fully or partly) to clear the wheat stubble used 169 kg of urea in the next crop of paddy while those who incorporated or adopted other means used 145 and 148 kg of urea, respectively. Similarly, those farmers who burnt paddy field, used added amount of Di-Amonia Phosphate (DAP) to recapture the nutritive lost in the fire in comparison to those who incorporated or removed stubble manually. Higher expenses were not only in terms of higher fertilizer but also in terms of higher irrigation requirement by those who burn their field to clear the stubble

Our household survey showed that paddy stubble burning leads to air pollution and several other problems. Irritation in eyes and congestion in the chest were the two major problems faced by the majority of the household members. Respiratory allergy, asthma and bronchial problems were the other smoke related diseases which affected household members in the selected villages. Almost 50 % of the selected households indicated that their health related problems get aggravated during or shortly after harvest when crop stubble burning is in full swing during the months of October, November and December. In the peak season, affected families had to consult doctor or use some home medicine to get relief from irritation/itching in eyes, breathing problem and similar other smoke related problems. On an average, the affected members suffered at least half a month from such problems and had to spend Rs. 300–500 per household on medicine. In addition there were few examples where a family member had to be hospitalized for 3–4 days and additional expenditure was incurred. On an average, households spent around more than a thousand Rupees on the non chronic respiratory diseases like coughing, difficulty in breathing, irregular heartbeat, itching in eyes decreased lung function etc., during the year 2008–2009. However, out of this total expenditure, around 40–50 % was spent during the months of October and November during the time of crop stubble burning. There was an additional cost in terms of household members remaining absent from work due to illness.

We find that total annual welfare loss in terms of health damages due to air pollution caused by the burning of paddy straw in rural Punjab amounts to Rs. 76 millions. These estimates could be much higher if expenses on averting activities, productivity loss due to illness, monetary value of discomfort and utility could also be considered. There is additional monetary cost of burning to the farmers in terms of additional fertilizer, pesticides and irrigation. One also needs to add the losses of soil nutrient, vegetation, bio-diversity and accidents caused because of low visibility.

Appendix

See Tables 3.26 and 3.27.

Table 3.26 Cropping pattern of selected farmers (percentage of gross cropped area)

Crop name	Marginal	Small	Medium	Large	Total
Wheat	40.3	27.4	25.2	25.2	28.5
Rice	33.8	27.4	25.2	25.2	27.3
Maize	0.0	0.0	0.0	0.9	0.2
Moong	0.0	0.0	0.0	0.9	0.2
Mustard	0.0	0.0	0.0	0.9	0.2
Sugarcane	0.0	0.0	0.0	0.9	0.2
Jowar and bajra (kharif green fodder)	16.9	23.0	25.2	23.5	22.6
Barseem (rabi green fodder)	9.1	22.1	24.3	22.6	20.4
Total	100.0	100.0	100.0	100.0	100.0

Table 3.27 Are there any buyers of rice/wheat residue

	Dhanori	Ajnauda Kalan	Simro	Aggregate
Are there any buyers for rice residue (percent of hh)	0.00	4.00	2.00	2.00
Are there any buyers for wheat residue (percent of hh)	10.00	16.00	10.00	12.00
Quantity of rice residue sold by households (quintals per hh)	0.00	2.50	5.60	2.80
Quantity of wheat residue sold by households (quintals per hh)	2.38	9.36	6.70	6.15
Average price of rice residue (Rs. per quintals)	10.00	10.00	10.00	10.00
Average price of wheat residue (Rs. per quintals)	200.00	200.00	200.00	200.00

References

Alberini, A., & Krupnick, A. (2000). Cost of illness and willingness to pay estimates of the benefits of improved air quality: Evidence from Taiwan. *Land Economics, 76*, 37–53.

Chesnut, L. G., Ostro, B. D., & Vichit-Vadakan, N. (1997). Transferability of air pollution control health benefits estimates from the United States to developing countries: Evidence from the Bangkok study. *American Journal of Agricultural Economics, 79*, 1630–1635.

Cropper, M., Simon, N. B., Alberini, A., Seema, A., & Sharma, P. K. (1997). The health benefits of air pollution control in Delhi. *American Journal of Agricultural Economics, 79*(5), 1625–1629.

Dockery, D. W., Pope, C. A., Xu, X., Spengler, J. D., Ware, J. H., Fay, M. E., et al. (1993). An association between air pollution and mortality in six U.S. cities. *New England Journal of Medicine, 329*, 1753–1759.

Gadde, B., Bonnet, S., Menke, C., & Garivait, S. (2009). Air pollutant emissions from rice straw open field burning in India, Thailand and the Philippines. *Environmental Pollution, 157*(5), 1554–1558.

Gerking, S., & Stanley, S. (1986). An economic analysis of air pollution and health: The case of St. Louis. *Review of Economics and Statistics, 68*(1), 115–121.

Gupta, U. (2008). Valuation of urban air pollution: A case study of Kanpur city in India. *Environmental and Resource Economics, 41*, 315–326.

Harrington, W., Krupnick, A. J., & Spofford, W. O. (1989). The economic losses of waterborne disease outbreak. *Journal of Urban Economics, 25*, 116–137.

Kumar, S., & Rao, D. N. (2001). Valuing benefits of air pollution abatement using health production function: A case study of Panipat Thermal Power Station, India. *Environmental and Resource Economics, 20*, 91–102.

Long, W., Tate, R. B., Neuman, M., Manfreda, J., Becker, A. B., & Anthonisen, N. R. (1998). Respiratory symptoms in a susceptible population due to burning of agricultural residue. *Chest, 113*, 351–357.

Murty, M. N., Gulati, S. C., & Banerjee, A. (2003). *Health benefits from urban air pollution abatement in the Indian subcontinent.* Discussion Paper No. 62/2003. Delhi: Institute of Economic Growth. www.ieg.org.

Ostro, B., Sanchez, J., Aranda, C., & Eskeland, G. S. (1995). *Air pollution and mortality: Results from Santiago, Chile.* Policy Research Department, Working Paper 1453. Washington, DC: World Bank.

Pope, C. A. 3rd., Thun, M. J., Namboodiri, M. M., Dockery, D. W., Evans, J. S., Spieizer, F. E., & Heath C. W. Jr. (1995). Particulate air pollution as a predictor of mortality in a perspective study of US adults. *American Journal of Respiratory and Critical Care Medicine, 151*(3), 669–674.

Punjab Pollution Control Board. (2007). Air pollution due to burning of crop residue in agriculture fields of Punjab. Assigned and Sponsored by PPCB, Patiala and CPCB, New Delhi. www.envirotechindia.com.

Schwartz, J. (1993). Particulate air pollution and chronic respiratory diseases. *Environmental Research, 62,* 7–13.

Chapter 4
Alternative Uses of Crop Stubble

Abstract Keeping in view the increasing problems associated with crop stubble burning in the state of Punjab, several initiatives for its proper management have been taken up. Various departments and institutions of the Punjab government are promoting alternative uses of straw instead of burning. This chapter outlines some of these alternative uses such as: use of rice residue as fodder; use of rice residue in bio-thermal power plants; its use for mushroom cultivation, for bedding material for cattle; its use for production of bio-oil; paper production; bio-gas and in situ. Other uses include incorporation of paddy straw in soil, energy technologies and thermal combustion.

Keywords Alternate uses of rice residue · End use of paddy straw · Residue use as fodder · Residue use in bio-thermal · In-situ incorporation

4.1 Introduction

Paddy straw is a major field-based residue that is produced in large amounts in Asia. In fact the total amount equaling 668 t could produce theoretically 187 gallons of bioethanol if the technology were available (Kim and Dale 2004). However, an increasing proportion of this paddy straw undergoes field burning. This waste of energy seems inapt, given the high fuel prices and the great demand for reducing greenhouse gas emissions as well as air pollution. As climate change is extensively recognized as a threat to development, there is a growing interest in alternative uses of field-based residues for energy applications.

Punjab produces around 23 million tonnes of paddy straw and 17 million tonnes of wheat straw annually. More than 80 % of paddy straw (18.4 million tonnes) and almost 50 % wheat straw (8.5 million tonnes) produced in the state is being burnt in fields. Almost whole of paddy straw, except Basmati rice is burnt in the field to enable early sowing of next crop. Lately, the farmers have extended this practice to wheat crop also. Though part of the wheat straw is used as dry fodder for the milch cattle, the remaining straw is usually burnt for quick disposal.

© The Author(s) 2015
P. Kumar et al., *Socioeconomic and Environmental Implications of Agricultural Residue Burning*, SpringerBriefs in Environmental Science,
DOI 10.1007/978-81-322-2014-5_4

There are primarily two types of residues from rice cultivation that have potential in terms of energy—straw and husk. Although the technology of using rice husk is well established in many Asian countries, paddy straw as of now is rarely used as a source of renewable energy. One of the principal reasons for the preferred use of husk is its easy procurement, i.e., it is available at the rice mills. In the case of paddy straw, however, its collection is a tedious task and its availability is limited to harvest time. The logistics of collection could be improved through baling but the necessary equipment is expensive and buying it is uneconomical for most rice farmers. Thus, technologies for energy use of straw must be efficient to compensate for the high costs involved in straw collection.

The chapter is organized as follows: The next section presents disposal pattern of paddy straw giving details of alternate uses of agriculture waste, viz., rice residue as fodder for animals, its use in bio-thermal power plants, its use for bedding material for animals, mushroom cultivation and so on. Section four discusses in details about residue use in power generation citing various biomass power projects commissioned in the state by Punjab Energy Development Agency (PEDA).

4.2 Disposal Pattern of Paddy Straw

The disposal pattern of paddy straw by the farmers depends on the market value of the by-product. Table 4.1 presents the methods adopted for end-use of paddy straw as mentioned in various studies. From the table, it is clear that on an average, three fourth of the paddy straw is burnt openly in the fields. The above ratio implies that in the year 2007–2008 around 11,930–15,858 thousand tonnes of paddy straw was burnt in the open field. Burning in Punjab involves partial and full burning. Partial burning entails running of combine harvester followed by burning of small stalks while complete burning entails setting the entire field on fire. The latter practice is mostly followed by the farmers in Punjab. Both the practices cause pollution

Table 4.1   End use of paddy straw

S. No	Author	Disposal pattern
1	Badarinath and Chand Kiran (2006)	75–80 % area is machine harvested
		¾ or 75 % of straw is burnt
2	Venkataraman et al. (2006)	30–40 % straw burnt (IGP)
3	Sidhu and Beri (2005)	81 % of paddy burnt and 48 % of wheat burnt, fodder (7 % of rice and 45 % of wheat), rope making (4 % of rice and 0 % of wheat), incorporated in soil (1 % of rice and less than 1 % of wheat), miscellaneous (7 % each of rice and wheat)
4	Sarkar et al. (1999)	75 % combine harvested and 100 % burnt
Average		75 % of paddy is burnt

Source Authors' compilation

but the impact is more severe in the case of complete burning. The farmers in the region are resorting to burning of straw, because they don't have other equal or more remunerative alternatives available to them.

There are many environmental risks associated with stubble burning. If followed continuously burning can reduce soil quality and make land more susceptible to erosion. Moreover, continuous burning is not a sustainable agricultural practice. Smoke from burning straw also contributes to increased carbon dioxide levels in the atmosphere which may affect greenhouse gas build-up.

The Department of Science, Technology and Environment and Non-Conventional Sources of Energy, Government of Punjab, constituted a task force in September, 2006 for formulation of policy to mitigate the problem due to the severity of burning of agricultural waste in the open fields after harvest and its consequent effects on soil, ambient air and health effects on living organism. The task force has suggested promotion of agronomic practices and technological measures for better utilization of agricultural wastes. These include use of happy seeder, developed by PAU in collaboration with Australian Centre for International Agriculture Research (ACIAR) and use of paddy straw for power generation.

4.3   Management of Agricultural Waste for Alternate Uses

Agricultural waste includes paddy and wheat straw, cotton sticks, bagasse and animal waste. Keeping in view the increasing problems associated with crop stubble burning several initiatives for its proper management have been taken up. Various departments and institutions are promoting alternative uses of straw instead of burning. These include:

4.3.1   Use of Rice Residue as Fodder for Animals

The rice residue as fodder for animals is not a very popular practice among farmers in Punjab.[1] This is mainly because of the high silica content in the rice residue. It is believed that almost 40 % of the wheat straw produced in the state is used as dry fodder for animals. However to encourage the use of rice residue as fodder for animals, a pilot project was taken up by PSCST at PAU under which trials on natural fermentation of paddy straw for use as protein enriched livestock feed were conducted. The cattle fed with this feed showed improvement in health and milk

---

[1] There are exceptions to this as in states like Kerala, the powder made out of the rice husk is fed (mixed with water) on to cattle so also the straw. Though it is reported to be unhealthy, probably the lack of other alternative sources of fodder compel people to use the same. It is also seen from Table 4.2, where the consumption of residue per animal is the highest at 0.35 t, (second to Punjab) which is much above many states. Rice being the main crop in Kerala, there is a high proportion of rice husk powder/straw consumption.

production. The technology was demonstrated in district Gurdaspur, Ludhiana, Hoshiarpur and Bathinda. The department of Animal Husbandry, Punjab has propagated the technology in the state. The analysis below presents the position of different states in production; availability and requirement of dry as well as green fodder and indicates which state is surplus/deficit in fodder requirement.

Table 4.2 indicates that total production of residue of paddy is almost 30 million tonnes for the total livestock of 464,472 thousands. Thus the consumption of paddy residue per livestock stands at 0.06 t/animal. Highest imbalance of livestock and consumption is noted in Rajasthan with zero consumption per animal. Other such low ranked state with least consumption rate is Madhya Pradesh,

Table 4.2 State-wise consumption of paddy (residue) per animal

States/UTs	Residue (000 tonnes)	Total Livestock (000)	Consumption of residue/animal
	TE 2006–2007	2003	(t/animal)
Andhra Pradesh	5,530	48,195	0.11
Arunchal Pradesh	71	1,261	0.06
Assam	1,657	13,431	0.12
Bihar	1,826	9,688	0.19
Chhattisgarh	2,406	13,487	0.18
Gujarat	654	21,168	0.03
Haryana	1601	8884	0.18
Himachal Pradesh	60	5,183	0.01
J & K	267	10,345	0.03
Jharkhand	1,034	15,478	0.07
Karnataka	2,123	25,621	0.08
Kerala	1,266	3,629	0.35
Madhya Pradesh	699	35,365	0.02
Maharashtra	1,236	35,770	0.03
Manipur	201	971	0.21
Meghalaya	91	1,552	0.06
Mizoram	39	280	0.14
Nagaland	131	1,349	0.10
Orissa	3,358	23,410	0.14
Punjab	5,128	8,608	0.60
Rajasthan	79	49,146	0.00
Sikkim	11	426	0.03
Tamil Nadu	2,482	24,126	0.10
Tripura	286	1,458	0.20
Uttar Pradesh	5,302	57,869	0.09
Uttaranchal	286	4,943	0.06
West Bengal	7,357	41,619	0.18
India	29,809	464,472	0.06

Source Lok Sabha Unstarred Question No. 726, dated on 24.11.2009

Himachal Pradesh, Maharashtra and Sikkim. In north, Punjab has got highest ratio of consumption, followed by Kerala and North Eastern state Tripura and Manipur. Uttar Pradesh has highest concentration of livestock which is followed by Rajasthan, Madhya Pradesh and Maharashtra. The residue is found highest in West Bengal and Arunachal Pradesh.

The availability of crop residue in India is 253.26 million tonnes whereas the requirement is 415.83 million tonnes (Table 4.3). Thus there is shortfall of almost 40 %. On the other hand, the availability of green fodder during the same time period is 142.82 million tonnes and requirement is 221.63 million tonnes with a short fall of almost 36 % (Table 4.4). It can be noted that only in Punjab and Mizoram there is surplus in case of crop residues.

Table 4.3 Status of different states about availability and requirement of fodder

States/UTs	Availability Crop residues (million tonnes)	Requirement	Livestock numbers (000)	Per animal availability (t/animal)	Per animal requirement (t/animal)
Andhra Pradesh	15.69	31.71	48,195	0.03	0.66
Arunachal Pradesh	0.47	1.00	1,261	0.04	0.79
Assam	5.82	12.39	13,431	0.04	0.92
Bihar	16.23	23.49	9,688	0.17	2.42
Chhattisgarh	9.93	14.93	13,487	0.07	1.11
Gujarat	10.61	22.32	21,168	0.05	1.05
Haryana	8.75	9.95	8,884	0.10	1.12
Himachal Pradesh	2.3	4.60	5,183	0.04	0.89
Jammu & Kashmir	2.53	6.79	10,345	0.02	0.66
Jharkhand	4.1	13.59	15,478	0.03	0.88
Karnataka	14.59	20.66	25,621	0.06	0.81
Kerala	0.71	2.91	3,629	0.02	0.80
Madhya Pradesh	24.3	37.41	35,365	0.07	1.06
Maharashtra	22.21	33.68	35,770	0.06	0.94
Manipur	0.36	0.72	971	0.04	0.74
Meghalaya	0.31	1.17	1,552	0.02	0.75
Mizoram	0.15	0.06	280	0.05	0.21
Nagaland	0.56	0.74	1,349	0.04	0.55
Orissa	12.25	22.27	23,410	0.05	0.95
Punjab	13.71	10.58	8,608	0.16	1.23
Rajasthan	21.67	33.53	49,146	0.04	0.68
Sikkim	0.23	0.25	426	0.05	0.59
Tamil Nadu	7.01	16.46	24,126	0.03	0.68
Tripura	0.53	1.09	1,458	0.04	0.75
Uttar Pradesh	42.07	57.19	57,869	0.07	0.99
Uttarakhand	2.05	4.90	4,943	0.04	0.99
West Bengal	13.77	30.30	41,619	0.03	0.73
India	253.26	415.83	464,472	0.05	0.90

Table 4.4 State-wise percentage of short fall of crop residue and greens

States/UTs	Avail-ability	Require-ment	Shortfall (%)	Availa-bility	Require-ment	Shortfall (%)
	Crop residues (million tonnes)			Green fodder (million tonnes)		
Andhra Pradesh	15.69	31.71	50.52	4.88	16.91	71.14
Arunachal Pradesh	0.47	1.00	53.00	1.57	0.53	−196.23
Assam	5.82	12.39	53.03	0.95	6.61	85.63
Bihar	16.23	23.49	30.91	0.81	12.53	93.54
Chhattisgarh	9.93	14.93	33.49	2.83	7.96	64.45
Goa	0.13	0.15	13.33	0.05	0.08	37.50
Gujarat	10.61	22.32	52.46	14.48	11.9	−21.68
Haryana	8.75	9.95	12.06	6.57	5.31	−23.73
Himachal Pradesh	2.3	4.60	50.00	1.98	2.45	19.18
J & K	2.53	6.79	62.74	0.64	3.62	82.32
Jharkhand	4.10	13.59	69.83	0.88	7.25	87.86
Karnataka	14.59	20.66	29.38	3.55	11.02	67.79
Kerala	0.71	2.91	75.60	0.39	1.55	74.84
Madhya Pradesh	24.3	37.41	35.04	11.65	19.95	41.60
Maharashtra	22.21	33.68	34.06	25.12	17.96	−39.87
Manipur	0.36	0.72	50.00	0.00	0.38	100.00
Meghalaya	0.31	1.17	73.50	0.4	0.62	35.48
Mizoram	0.15	0.06	−150.00	0.5	0.03	−1,566.67
Nagaland	0.56	0.74	24.32	0.3	0.4	25.00
Orissa	12.25	22.27	44.99	2.46	11.88	79.29
Punjab	*13.71*	*10.58*	*−29.58*	*7.38*	*5.64*	*−30.85*
Rajasthan	21.67	33.53	35.37	33.53	17.88	−87.53
Sikkim	0.23	0.25	8.00	0.01	0.13	92.31
Tamil Nadu	7.01	16.46	57.41	3.7	8.78	57.86
Tripura	0.53	1.09	51.38	0.19	0.58	67.24
Uttar Pradesh	42.07	57.19	26.44	15.73	30.5	48.43
Uttarakhand	2.05	4.90	58.16	1.73	2.61	33.72
West Bengal	13.77	30.30	54.55	0.51	16.16	96.84
A & N Islands	0.02	0.11	81.82	0.00	0.06	100.00
Chandigarh	0.00	0.04	100.00	0.00	0.02	100.00
Dadra & Nagar H	0.04	0.80	95.00	0.20	0.40	50.00
Daman Diu	0.01	0.10	90.00	0.00	0.00	–
Delhi	0.09	0.43	79.07	0.10	0.23	56.52
Lakshadweep	0.00	0.10	100.00	0.00	0.00	–
Pondicherry	0.06	0.11	45.45	0.01	0.06	83.33
India	*253.26*	*415.83*	*39.10*	*142.82*	*221.63*	*35.56*

Source Lok Sabha Unstarred Question No. 726, dated on 24.11.2009

The availability of crop residue is highest in Uttar Pradesh followed by Maharashtra, Bihar, Rajasthan and Andhra Pradesh. Excepting Assam almost all the north Eastern States and Kerala have least availability of crop residue. As in the case of availability, the highest requirement of crop residue is in Uttar Pradesh and thus the requirement per animal (0.99 t/animal) and per animal availability of the state is also high (0.07 t/animal). States like Punjab, Haryana and Bihar has higher per animal availability as compared to other states of India.

4.3.2 Use of Crop Residue in Bio Thermal Power Plants

Another use of rice residue that is being encouraged by various institutions and departments is the use of rice residue for generation of electricity. A 10 MW bio-mass based power plant at village Jalkheri, Fatehgarh Sahib with paddy straw as fuel was set up in the year 1992 (Box 4.1). The plant is operational since 2001, after the PSEB entered into a lease-cum-power purchase agreement with Jalkheri Power Private Limited (JPPL). The original system installed by BHEL i.e. firing the boiler with paddy straw in baled form, used to create innumerable problems like ash melting, snagging, super heater choking, clinkerisation, drop in boiler temperature due to moisture in the bales, etc. Hence, the fuel was changed from paddy straw to rice husk, wood chips, cotton waste, etc., in mixed form or rice husk alone to achieve the desired parameters. The total requirement of biomass is estimated to be 82,500 MT/annum at 100 % capacity utilization for optimum plant activity. Crop residues are bought from the farmers at Rs. 35 per quintal (which would otherwise have remained unutilized or burnt in the field). The farmers are being made aware of this offer through newspapers and other awareness activities. Apart from the generation of electricity for supply to state grid to meet the ever-increasing demand for energy in the state, the plant also reduces the Green House Gases (GHGs) emissions. As per Cleaner Development Mechanism (CDM) estimates, the plant would supply energy equivalent of approximately 417.9 million kWh to the grid in a period of 10 years (2002–2012), thereby resulting in total CO_2 emission reduction of 0.3 million tonnes.

Box 4.1 Case Study of Generation of Electricity from Agri-Waste

The thermal plant at Jalkheri, District Fatehgarh Sahib is the first plant in India which is based on use of Biomass i.e. renewable energy source. This plant can utilize rice husk, waste wood chips, straw of various plants e.g. paddy, wheat, etc. This plant was commissioned in June, 1992 on turn-key basis by M/s BHEL for PSEB to utilize rice straw at a cost of Rs. 47.2 crores.

Some teething problems were experienced initially being an experimental project, but with modifications, full 10 MW capacity has been achieved. As harvesting pattern in Punjab has changed and farmers found it convenient to harvest crop with mechanical means and non-availability of adequate quantity of hand cut rice, the plant was further modified to accept any bio-mass e.g. any straw, rice husk, wood chips etc. The plant has been given on lease and is being operated at 10 MW i.e. full capacity on sustainable basis.

One 10–15 MW agri-waste based power project has been set up jointly by Punjab Biomass Power, Bermaco Energy, Archean Granites and Gammon Infrastructure projects Limited in Punjab. The project uses locally available agricultural waste such as rice straw and sugar cane trash for fuel. The total annual fuel requirement is around 120,000 t of biomass, all of which will be sourced locally. Punjab produces around 20–25 million tonnes of rice straw annually. As rice straw is a poor fodder and fuel, farmers burn it in the fields and make way for the Rabi wheat crop. With the development of technology now there is an option to use this waste for generating electricity. The project is expected to provide additional income to 15,000 farmers from the sale of agricultural waste. The project will be a major milestone in environment protection—converting agricultural waste to energy. Secondly, it will reduce the release of smoke and other pollutants caused by burning of wastes which could now be used for earning carbon credits.

Another biomass based power project of 7.5 MW was initiated by Malwa Power Pvt. Ltd. at village Gulabewalla in district Mukatsar in 2002. The project was commissioned in May 2005 and is operating satisfactorily. The plant is selling electricity to PSEB through power purchase agreement. The plant is using crop residues available in the area like cotton stalks, mustard stalks, lops and tops of Eucalyptus, Poplar and Prosopis juliflora and some quantity of agro waste such as rice husk and saw dust. The total requirement of biomass is estimated to be 65,043 MT per annum at 90 % capacity utilization and 72,270 MT per annum at 100 % capacity utilization.

As per estimates for Clean Development Mechanism (CDM), this plant would supply energy equivalent of approximately 465.10 million kWh to the grid in a period of 10 years (2005–2015) and would result in reduction of 0.43 million tonnes total of CO_2 emission. Both these power plants are obtaining Carbon Credits under CDM. Further, in August, 2006, PSEB has signed two agreements with M/S Punjab Biomass Power Limited for setting up 12 MW paddy straw based power plants at village Baghaura near Rajpura and Village Sawai Singh near Patiala. The company intends to collect paddy straw from command area of 25 km^2 around each village and would use 1 lakh MT per annum paddy straw for generation of 12 MW of electricity. The company has entered into an agreement with farmers on barter system and farmers will be provided electricity in lieu of supplying paddy straw. The plants were expected to start operations in 2009. Land at Baghaura village has already been purchased.

4.3.3 Use of Rice Residue as Bedding Material for Cattle

The farmers of the state have been advised to use paddy straw as bedding material for cross bred cows during winters as per results of a study conducted by the Department of Livestock Production and Management, College of Veterinary Sciences, Punjab Agricultural University. It has been found that the use of paddy straw bedding during winter helped in improving the quality and quantity of milk as it contributed to animals' comfort, udder health and leg health. Paddy straw bedding helped the animals keep themselves warm and maintain reasonable rates of heat loss from the body. It also provides clean, hygienic, dry, comfortable and non-slippery environment, which prevents the chances of injury and lameness. Healthy legs and hooves ensure enhancement of milk production and reproductive efficiency of animals. The paddy straw used for bedding could be subsequently used in biogas plants. The use of paddy straw was also found to result in increased net profit of Rs. 188–971 per animal per month from the sale of additional amount of milk produced by cows provided with bedding. The PAU has been demonstrating this technology to farmers through training courses, radio/TV talks and by distributing leaflets.

4.3.4 Use of Crop Residue for Mushroom Cultivation

Paddy straw can be used for the cultivation of Agaricus bisporus, *Volvariella Volvacea* and Pleurotus spp. One kg of paddy straw yields 300, 120–150 and 600 g of these mushrooms, respectively. At present, about 20,000 metric tonnes of straw is being used for cultivation of mushrooms in the state.

Paddy Straw Mushrooms (*Volvariella Volvacea*) also known as grass mushrooms are so named for their cultivation on paddy straw used in South Asia. Paddy Straw is high temperature mushroom grown largely in tropical and subtropical regions of Asia, e.g. China, Taiwan, Thailand, Indonesia, India, and Madagascar. In Indonesia and Malaysia, mushroom growers just leave thoroughly moistened paddy straw under trees and wait for harvest. This mushroom can be grown on a variety of agricultural wastes (the cultivation method of this mushroom is similar to that of Agaricus bisporus) for preparation of the substrate such as water hyacinth, oil palm bunch waste, dried banana leaves, cotton or wood waste, though with lower yield than with paddy straw, which is most successful. Paddy straw mushroom accounts for 16 % of total production of cultivated mushroom in the world.

4.3.5 Use of Rice Residue in Paper Production

The paddy straw is also being used in conjunction with wheat straw in 40:60 ratios for paper production. The sludge can be subjected to bio-methanization for energy production. The technology is already operational in some paper mills, which are meeting 60 % of their energy requirement through this method. Paddy straw is also used as an ideal raw material for paper and pulp board manufacturing. As per information provided by PAU, more than 50 % pulp board mills are using paddy straw as their raw material.

4.3.6 Use of Rice Residue for Making Bio Gas

The PSFC has been coordinating a project for processing of farm residue into biogas based on the technology developed by Sardar Patel Renewable Energy Research Institute (SPRERI). A power plant of 1 MW is proposed to be set up at Ladhowal on pilot basis on land provided by PAU. The new technology will generate 300 m^3 of biogas from 1 t of paddy straw.

4.3.7 In Situ

The technical measures are 'straw incorporation' and 'straw mulching'. In both these measures, the residue is incorporated in the field itself and is thus used to increase the nutrient value or fertility of the soil. In the first measure, the residue is allowed to decompose in the field itself through a chemically developed process (available at PAU), and in the second measure, incorporation is done with the help of a properly designed machine along with seeding (know-how developed at PAU). The second measure is more useful as there is no weeding in this process and it is less expensive.

Another study (Singh 1992) reveals that, incorporation of paddy straw in soil immobilized native as well as added fertilizer N and about half of the immobilized N was mineralized after 90 days of straw incorporation. Straw and N application alone or in combination increased biomass carbon, phosphates and respiratory activities of the soil. Microbial biomass carbon and phosphate activities were observed maximum at 30 days of straw decomposition. In field trials, incorporation of paddy straw 3 weeks before sowing of wheat significantly increased the wheat yield at Sonepat district in a clay loam soil while no such beneficial effect was observed in a sandy loam soil at Hissar (Singh 1992).

4.3.8 Incorporation of Paddy Straw in Soil

The incorporation of the straw in the soil has a favorable effect on the soil's physical, chemical and biological properties such as pH, Organic carbon, water holding capacity and bulk density of the soil. On a long-term basis it has been seen to increase the availability of zinc, copper, iron and manganese content in the soil and it also prevents the leaching of nitrates. By increasing organic carbon it increases bacteria and fungi in the soil. In a rice-wheat rotation, Beri et al. (1992) and Sidhu et al. (1995) observed that soil treated with crop residues held 5–10 times more aerobic bacteria and 1.5–11 times more fungi than soil from which residues were either burnt or removed. Due to increase in microbial population, the activity of soil enzymes responsible for conversion of unavailable to available form of nutrients also increases. Mulching with paddy straw has been shown to have a favorable effect on the yield of maize, soybean and sugarcane crops. It also results in substantial savings in irrigation and fertilizers. It is reported to add 36 kg per hectare of nitrogen and 4.8 kg per hectare of phosphorous (6 g of Nitrogen and 0.8 g of phosphorous per kg of paddy straw) leading to savings of 15–20 % of total fertilizer use.

4.3.9 Production of Bio-oil from Straw and Other Agricultural Wastes

Bio-oil is a high density liquid obtained from biomass through rapid pyrolysis technology. It has a heating value of approximately 55 % as compared to diesel. It can be stored, pumped and transported like petroleum based product and can be combusted directly in boilers, gas turbines and slow and medium speed diesels for heat and power applications, including transportation. Further, bio-oil is free from SO_2 emissions and produces low NO_2. Certain Canadian companies (like Dyna Motive Canada Inc.) have patented technologies to produce bio-oil from biomass including agricultural waste. Though their major experience is with bagasse, wheat straw and rice hulls, feasibility of this technology with paddy straw needs to be assessed. The state government, through PSCST and PEDA, could promote further studies in this direction.

4.4 Agricultural Residues for Power Generation

The State of Punjab has been a victim of acute power famines, load shedding and power cuts, year after year. Agricultural requirement for power is highest during June to September for the purpose of paddy cultivation. Biomass, such as agricultural residue, bagasse, cotton stalks, rice husk, etc., is emerging as a viable source of power for rural electrification in India. Direct burning of such waste is inefficient and leads to pollution. When combusted in a gasifier at low oxygen and high temperature, biomass can be converted into a gaseous fuel known as producer gas. This gas has a lower calorific value compared to natural gas or liquefied petroleum gas, but can be burnt with high efficiency and without emitting smoke.

The advantages of utilizing crop residue over and above the conventional resources are that such residue is renewable, readily available and can be used successfully by burning in boilers with the efficiency of 99 %. Further, they are available at low cost as compared to that of coal while ash contents is much less (as compared to 36 % ash content of coal) and at the same time the calorific value of both, coal and paddy straw are comparable, i.e., 4,200 and 3,590 kcal/kg, respectively. Additional income to the farmers from the sale of straw is an added advantage. At the same time, the agencies involved/state could also take advantage of carbon credit policy set up under the UNFCCC (United Nation Framework Convention on Climate Change) from developed countries. The policy involves emission credit for programmes which help in curbing global warming. The government should encourage private parties/agencies to take advantage of this carbon credit policy of UNFCCC.

According to Dr. A.K. Rajvanshi, who runs the non-profit Nimbkar Agriculture Research Institute, Phaltan, Maharashtra, it is feasible to set up a bio-mass-based power plant of 10–20 MW capacity in every Taluka (a block of about 100 villages). This can meet energy needs of villages and employ thousands of people. Similarly, in Punjab the developers of biomass energy can sell their power to PSEB, which will be purchased as per, 'New & Renewable Sources of Energy Policy' notified by the government from time to time and distributed as per usual norms.

Kirangatevalu village in Karnataka has set an example in this regard. Electrification of the village earlier meant supply of power to a few homes and farms for 4–5 h a day. The transformation of the village is the result of an initiative taken by a private firm that has set up a power plant using agricultural waste such as sugarcane refuse and coconut fronds that are plentiful in the area. Villagers sell their agro waste to the plant and get access to quality power at commercial rate. A supply chain to procure agricultural waste from villages in a radius of 10 km has been established to ensure the supply of agricultural waste throughout the year. The waste that was burnt in open fields has now become a source of income and jobs. The 4.5 MW power plants set up by Malaballi Power Plant Private Limited supplies electricity to 48 villages inhabited by 120,000 people in Mandya district in Karnataka.

In Punjab in the 1980s PSEB had set up a 10 MW power plant based on paddy straw at Village Jalkheri, District Fatehgarh Sahib in which 250–3,000 TPD of fuel is burnt in a boiler furnace of steam generation capacity of 50 TPH. The plant earlier used paddy straw but due to clinkerisation of boiler, paddy straw was replaced with rice husk, cow dung and other agro waste. This plant has since been leased out by PSEB to M/S Jalkheri Power Private Limited. Now these plants will be using improved technology and M/S Punjab Biomass Power Limited has signed two agreements with PSEB for setting up 12 MW paddy straw based power plants at Baghaura in Rajpura Tehsil and Sawai Singh village in Patiala Tehsil. A total amount of 0.1 million ton paddy straw would be collected from a command area of 25 km^2 around each unit and a barter system of providing electricity will be worked out with the farmers. The units will be run on BOO basis. DPRs have been prepared and land is being purchased.

The bottlenecks apprehended by PSEB in generation of power from paddy straw are the availability of paddy straw for power generation in case the Happy

Seeder technology succeeds in the State. Hence, it is recommended that areas around these power plants could be reserved to ensure enough availability of straw. Further, techniques to collect and store paddy straw may also be developed and incentives provided.

4.4.1 Energy Technologies

The transportation of biomass is one of the key cost factors for its use as a source of renewable energy. Decentralized energy systems provide an opportunity to use biomass to meet local energy requirements that are, heat and electricity. In contrast to straw, the use of rice husk for energy has been realized faster. One important factor is that rice mills can use husk to serve their internal energy requirement. As an alternative, rice millers could sell the husk to a power-plant operator. The propagation of rice husk use for energy was accelerated by energy providers, who deal with a relatively small number of rice millers for supplying husk, which is an easier task than dealing with thousands of farmers supplying paddy straw.

As a new trend, electricity is now often produced by the millers themselves and then sold to a power grid. This setup has to be seen as the most promising option in terms of logistics and transportation for energy generation. Transportation costs of straw are a major constraint to its use as an energy source. As a rule of thumb, transportation distances beyond a 25–50 km radius (depending on local infrastructure) are uneconomical. For long distances, straw could be compressed as bales or briquettes in the field, rendering transport to the site of use a viable option. Nevertheless, the logistics of a supply chain is more complicated in the case of straw.

Although five different energy conversion technologies seem to be applicable for paddy straw in principle only combustion technology is currently commercialized and the other technologies are at different stages of development. As a general rule for energy use, each step in the chain consumes a certain amount of energy and thus reduces the net energy at the end product. The following sections describe the principal features of the possible energy conversion technologies, experiences and technical difficulties in the use of paddy straw.

4.4.2 Thermal Combustion

Paddy straw can either be used alone or mixed with other biomass materials (the latter is called co-firing or co-combustion) in direct combustion. In this technology, combustion boilers are used in combination with steam turbines to produce electricity and heat. In thermal combustion, air is injected into the combustion chamber to ensure that the biomass is completely burned in the combustion chamber. Fluidized bed technology is one of the direct combustion techniques in which solid fuel is burned in suspension by forced air supply into the combustion chamber to achieve complete combustion. A proper air-to-fuel ratio is maintained and, in the absence of a sufficient air supply, boiler operation encounters various problems.

In straw combustion at high temperatures, potassium is transformed and combines with other alkali earth materials such as calcium. This in turn reacts with silicates, leading to the formation of tightly sintered structures on the grates and at the furnace wall. Alkali earths are also important in the formation of slags and deposits. This means that fuels with lower alkali content are less problematic when fired in a boiler (Jenkins et al. 1998). The byproducts are fly ash and bottom ash, which have an economic value and could be used in cement and/or brick manufacturing, construction of roads and embankments, etc.

National Biomass Assessment Project of Ministry of New and Renewable Energy, Government of India conducted a biomass study in which 29 Tehsil were surveyed which was started in the late eighties and continued till 1995–1996. Total 36 Talukas were included from different districts. The total estimated power generating potential was estimated AT 342 MW. Biomass Power project has the following inherent advantages over thermal power generation:

- It is environmentally friendly because of relatively lower CO_2 and particulate emissions
- It displaces fossil fuels such as coal
- It is a decentralised, load based means of generation, because it is produced and consumed locally, losses associated with transmission and distribution are reduced
- It offers employment opportunities to locals
- It has a low gestation period and low capital investment
- It helps in local revenue generation and upliftment of the rural population
- It is an established and commercially viable technology option.
- Punjab has substantial availability of Biomass/Agro-waste in the state sufficient to produce about 1,000 MW of electricity.

PEDA has planned to develop some of the available potential talukas/tehsils with the aim to promote and install biomass/agro waste based projects. PEDA has so far allocated 30 sites/tehsils for setting up of total 332.5 MW capacity Biomass/Agro waste based power projects under three phases. In different phases the biomass power project were allocated.

- In Phase I agreement is already done with two companies- M/s Turbo Atom TPS and M/S. Green Field Energen Pvt. Ltd., in New Delhi and Chandigarh, respectively for two Tehsils, Ferozepur and Patti with a total capacity of 56 MW.
- In Phase II three companies were there for Abohar, Sunam and Ajnala, respectively. The two companies of Sunam and Ajnala are cancelled having 41 MW. With the capacity of 8 MW the company M/s Dee Development in Abohar Tehsil is commissioned.

- In Phase III there are six companies which are based differently. The M/s Green Planet of Chandigarh is based on paddy stubble which is planned in 14 Tehsils with total 146.5 MW of capacity. Out of which Garhshankar with 10 MW capacity is likely to begin. The M/s Univeral Biomass of Mukatsar which is mostly based on cotton stock with 14.5 MW in Malout Tehsil is commissioned. The Malwa Power Ltd., in the village Gulabevala in the district of Muketsar was started before PEDA took over with 6 MW. Other three companies had total capacity of 65 MW.

Thus, PEDA has so far allocated 30 sites/tehsils for setting up of total 332.5 MW capacity Biomass/Agro-waste based power projects during three phases (details in Appendix).

4.5 Summary of the Chapter

To avoid burning of rice (and wheat) stubble, management of agricultural waste for alternate uses is being practiced and promoted. Agricultural waste includes paddy and wheat straw, cotton sticks, bagasse and animal waste. Keeping in view the increasing problems associated with crop stubble burning several initiatives for its proper management have been taken up. Various departments and institutions are promoting alternative uses of straw instead of burning. These include use of rice residue as fodder, crop residue in Bio thermal power plants and mushroom cultivation, rice residue used as bedding material for cattle, production of bio-oil, paper production, bio-gas and in situ. Other uses include incorporation of paddy straw in soil, energy technologies and thermal combustion. Although five different energy conversion technologies seem to be applicable for rice straw in principle only combustion technology is currently commercialized and the other technologies are at different stages of development. PEDA has so far allocated 30 sites/tehsils for setting up of total 332.5 MW capacity Biomass/Agro-waste based power projects during three phases.

Box 4.2 Punjab Farmers Take Lessons on Straw Management, Swarleen Kaur, Posted: Thursday, Feb 04, 2010 at 2,253 h IST, Financial Express

Chandigarh: Punjab government, an entrepreneur and an NGO have joined hands to bring about a change in the way farming is done in the state. To fight with the problem of burning paddy straw in fields, farmers are being given lessons and field training on rice straw management. Farmers are taught eco friendly way of zero tillage and how to use straw as an organic fertilizer. The ban imposed by state government on burning paddy residue meant little for growers and they continued to set fire to the dry straw in the state.

On an average, paddy is grown in around seven million acres in the state. An acre yields around 25–30 quintals of crop residue, thereby the aggregate crop residue is estimated at 175 million quintal of which more than 90 % paddy straw is burnt. It is estimated that farmers burn 19.6 million tonnes straw every year that is worth crores of Rupees, besides losing 38.5 lakh tonne of organic carbon, 59,000 t nitrogen, 2,000 t phosphorous and 34,000 t potassium every year. "We have decided to educate farmers of Ferozepur district about the adverse effects of residue burning. By focusing on better straw management, farmers can cut down their input costs, save water, fuel, use organic matter, make additional money while nursing the environment at the same time. Initially our team will provide training to Sarpanches who act as opinion leaders in villages. We will promote the use a new post-harvest-technology machine happy seeder that helps farmers in the incorporation of rice crop residue", Vikram Ahuja who runs Zamindara Farm Solutions, a farm equipment bank in Fazilika told FE.

Contrary to the local belief that rice straw is not a very good cattle fodder, farmers will be educated by taking them to cowsheds. Properly cut, chopped, collected and baled straw can also be sold at profitable price, he highlighted. Cereal Systems Initiative, a non-governmental initiative for South Asia (CSISA) headed by HS Sidhu in Punjab, has volunteered to offer technology to one part of this campaign. Elaborating on the concept, Sidhu said, "Under the CSISA project we have carried out eight sessions with farmers and have given 150 demonstrations. It has been found that farmers are able to save Rs. 1,500–1,800 if they use scientific methods. We intend to cover Amritsar, Kapurthala, Ludhiana, Fatehgarh-sahib, Patiala, Sangrur and Bathinda. This programme will be expanded gradually". According to agri-experts, only 15 % of the total paddy straw being produced in Punjab can be used in a productive way. PS Rangi, consultant with Punjab State Farmers' Commission (PSFC) told FE, "In Punjab, October onwards there is a haze over the countryside since paddy residue, being moisture and silica rich, keeps burning for days. This residue cannot be ploughed back either and since it is rich in silica, decomposition takes a long time. In the absence of viable alternatives, farmers are left with no other option but to burn paddy stubble. In such a situation the Farmers Commission as well as the agricultural department are promoting rotavators, which buries the paddy straw in the fields and happy seeders that uses the zero tilling method to sow wheat in the fields with paddy straw given that only 10 % of total paddy straw can be used to produce electricity at bio-mass power projects".

Appendix

A. Biomass Power Projects Commissioned in the State by PEDA: (52.5 MW)
(*Source*http://peda.gov.in/eng/Bio-mass%20Power.html; *accessed on 25 May 2014*)

S. No.	Name of the company	SITE	CAPCITY (MW)	Month of commissioning	Remarks
01	M/s Malwa Power Ltd.	Vill. Gulabewala, Distt. Mukatsar	6	May 2005	First project allocated by PEDA
02	M/s Dee Development Engineers Pvt. Ltd.	Vill.GaddaDhob, Tehsil. Abohar Distt Ferozepur	8	Feb 2009	Project was allocated under Phase-II
03	M/s Universal Biomass Energy Pvt. Ltd.	Vill. ChannuTeh. Malout, Distt. Sri Mukatsar Sahib	14.5	Oct 2009	Project was allocated under Phase-III
04	M/s. Punjab Biomass Power Pvt. Ltd.	Distt. Patiala	12	June 2010	Project allocated by PSPCL
05	M/s. Green Planet Energy Pvt. Ltd.	Binjon, Distt. Hoshiarpur	6	March 2012	Project was allocated under Phase-III
06	M/s. Green Planet Energy Pvt. Ltd.	Bir Pind, Distt. Jallandhar	6	Feb 2013	Project was allocated under Phase-III
Total			52.5		

B. Detailed status of project work of biomass power projects being setup by private developers allocated by PEDA

Total no. of sites initially allocated	31–348 MW
Total no. of sites—projects commissioned	5–40.5 MW
Total no. of sites cancelled so far	13–142 MW
Total no. of balance sites	13 + 2–4 MW
Total capacity	165.5 MW
Phase I	20 MW (2 Nos.)
Phase II	One project of 8 MW commissioned
Phase III	145.5 MW (11 Nos)

C. Company wise-status report of biomass power projects

1. M/s. Green Planet Energy Pvt. Ltd.

S. No.	Name of site	Capacity (MW)	Project status		Scheduled date of commissioning
			Activities completed	Activities in process	
1	Vill. Binjon, Tehsil Garhshankar, Distt. Hoshiarpur	6 + 4	6 MW Rankine Cycle:- project commissioned in May 2012	90 % Mechanical and Electrical works of 4 MW project is completed. 100 %Civil works completed	4 MW—June 2013
2	Vill. BirPind, Tehsil Nakodar, Distt. Jalandhar	6 + 4	6 MW Rankine Cycle:- project commissioned in Feb. 2013	4 MW Otto cycle:- civil works 30 % completed, G. Engine reached at site	4 MW—June 2014
3	Vill. Manuke Gill, Tehsil Nihal Singh Wala,,Distt. Moga	6	Land acquired MoU, IA and PPA signed	100 % Civil works completed. Erection of transmission line under process	6 MW—May 2013
4	Vill. Ramiana, Tehsil Jaito, Distt. Faridkot	12 + 1	Land acquired MoU, IA and PPA signed	Civil construction work yet to start. Order placed for Boilerand turbine	12 MW—Dec 2014 1 MW—Dec 2014
5	Vill. Deep Singh Wala, Tehsil & Distt. Faridkot	6 + 3	Land acquired MoU, IA and PPA signed	Civil construction work yet to start	6 MW—Sept 2014 3 MW—March 2015
6	Vill. TalwandiRai, Tehsil Raikot, Distt. Ludhiana	12 + 2	Land acquired MoU, IA and PPA signed	Civil construction work yet to start	12 MW—Dec 2014 2 MW—March 2015
7	V. Dhanasu Teh & Distt. Ludhiana	15	Land acquired MoU signed	Company has taken the Panchayat Land on lease	March 2015
8	V. Borana Teh & Distt. Fatehgarh Sahib	12	Land acquired MoU signed IA & PPA not signed	Company has taken the Panchayat Land on lease. Civil construction work yet to start	March 2015
9	Tehsil Bathinda	13.5	MOU signed	Land not taken	Dec 2015
	Total	*102.5*			

2. M/s. Turboatom-TPS Projects Pvt. Ltd.

S. No.	Name of site	Capacity (MW)	Project status		Scheduled date of commissioning
			Activities completed	Activities in process	
1	Vill. BurjBaghel Singh, Malerkotla (Sangrur)	20	MoU signed Land acquired IA signed on dated 13.12.2010 PPA signed on 10.6.2011	No work started	Dec 2013
2	Vill. JhokTehal Singh, Ferozepur (Ferozepur)	10	Land acquired MoU signed IA signed PPA with PSPCL signed	No work started	Dec 2013
	Total	*30*			

3. M/s. Orient Green Power Pvt. Ltd.

S. No.	Name of site	Capacity (MW)	Project status		Scheduled date of commissioning
			Activities completed	Activities in process	
1	Vill. WadalaBhittiwind Teh. Amritsar	10	Land acquired MoU signed	Company asked to sign IA with PEDA & PPA with PSPCL Civil works yet to start at site	Sept 2014
	Total	*10*			

4. M/s. Viaton Energy Pvt. Ltd. (formerly M/s. Food Fats & Fertilizers Pvt. Ltd.)

S. No.	Name of Site	Capacity (MW)	Project status		Scheduled date of Commissioning
			Activities completed	Activities in process	
1	Vill. KhokharKhurd, Tehsil Mansa	20	Land acquired MoU signed IA & PPA signed	Site mobilized, 70 % civil works completed, boiler erection in progress, other boiler components already reached site. Turbine already imported and reaching site shortly	March 2013
	Total	*20*			

5. M/s. P & R Agri Energy Pvt. Ltd.

S. No.	Name of site	Capacity (MW)	Project status		Scheduled date of commissioning
			Activities completed	Activities in process	
1	Vill. Gopalpur Teh. Anandpur Sahib	5	Land acquired MoU signed IA signed on 10th Nov.'2010 PPA signed 8.8.2011	Civil work started Work orders placed for boiler and turbine	Sept 2014
2	Vill. PatharmajraTeh. Ropar	10	MoU signed Land acquired	Company asked to sign IA with PEDA & PPA with PSPCL Company asked to start the project work immediately	Sept 2014
	Total	*15*			

References

Badarinath, K. V. S., & Chand Kiran, T. R. (2006). Agriculture crop residue burning in the Indo-Gangetic Plains—A study using IRSP6 WiFS satellite data. *Current Science, 91*(8), 1085–1089.

Beri, V., Sidhu, B. S., Bhat, A. K., & Singh, B. P. (1992). Nutrient balance and soil properties as affected by management of crop residues. In: M.S. Bajwa et. al. (Eds.), *Nutrient management for sustained productivity* (pp. 133–135). Proceedings of International Symposium (vol. II). Ludhiana, India: Department of Soil, Punjab Agricultural University.

Jenkins, B. M., Baxter, L. L., Miles, T. R, Jr, & Miles, T. R. (1998). Combustion properties of biomass. *Fuel Processing Technology, 54*, 17–46.

Kim, S., & Dale, B. E. (2004). Cumulative energy and global warming impacts from the production of biomass for biobased products. *Journal of Industrial Ecology, 7*(3–4), 147–162.

Sarkar, A., Yadav, R. L., Gangwar, B., & Bhatia, P. C. (1999). *Crop residues in India.* Modipuram: Project Directorate for Cropping System Research. Tech. Bull.

Sidhu, B. S., & Beri, V. (2005). Experience with managing rice residues in intensive rice-wheat cropping system in Punjab. In I. P. Abrol, R. K. Gupta, & R. K. Malik (Eds.), *Conservation agriculture: Status and prospects* (pp. 55–63). New Delhi: Centre for Advancement of Sustainable Agriculture, National Agriculture Science Centre.

Sidhu, B. S., Beri, V. & Gosal, S. K. (1995) Soil microbial health as affected by crop residue management. In *Proceedings of National Symposium on Developments in Soil Science, Ludhiana, India* (pp. 45–46). New Delhi, India: Indian Society of Soil Science. 2–5 November, 1995.

Singh, S., Batra Renu, Mishra, M. M., Kapoor, K. K., & Goyal Sneh (1992). Decomposition of paddy straw in soil and the effect of straw incorporation in the field on the yield of wheat. *Journal of Plant Nutrition and Soil Sciences, 155*(4), 307–311.

Venkataraman, C., Habib, G., Kadamba, D., Shrivastava, M., Leon, J.F., Crouzille, B., Boucher O., & Streets D. G. (2006). Emissions from open biomass burning in India: Integrating the inventory approach with high-resolution Moderate Resolution Imaging Spectroradiometer (MODIS) active-fire and land cover data. *Global Biogeochemical Cycles 20*(2), 1–12.

Chapter 5
Environmental Legislations: India and Punjab

Abstract India is a legislation rich country with reference to pollution. Eleven major laws exist to control pollution in India and many forums for their implementation in various ways. Under these laws, provisions are made to protect the environment from all kinds of pollution related to industrial and agricultural activities. The Punjab Pollution Control Board (PPCB) is entrusted with the functions of planning a comprehensive program for the prevention, control and abatement of pollution in Punjab. PPCB has to support and encourage developments in the field of pollution control. PPCB has taken various measures to limit the amount of industrial pollution in the state but not much has been done to address agricultural pollution (http://www.ppcb.gov.in/index.aspx).

Keywords Legislation to control pollution · Central Pollution Control Board · Punjab Pollution Control Board · Punjab Energy Development Agency

5.1 Introduction

This chapter discusses the legislation on pollution in India in general and Punjab in particular. It presents provisions of various laws to control pollution like Water Act 1974, Air Prevention and Control of Pollution Act 1981, Environment Protection Act 1986, National Environment Tribunal Act 1995, Noise Pollution Rules 2000, Bio-diversity Act 2000 and so on. The chapter also discusses various functions and activities of Central Pollution Control Board, Punjab Pollution Control Board, Punjab State Council for Science and Technology, Punjab Energy Development Agency and Punjab Bio Diversity Board to control various types of pollution (http://www.pscst.gov.in/).

© The Author(s) 2015
P. Kumar et al., *Socioeconomic and Environmental Implications of Agricultural Residue Burning*, SpringerBriefs in Environmental Science,
DOI 10.1007/978-81-322-2014-5_5

5.1.1 Ministry of Environment and Forest

The Ministry of Environment and Forest (MoEF) is a nodal agency in the administrative setup of the Union Government. The Ministry is entrusted with the task of planning, coordinating, overseeing and implementing various forestry and environment programmes. The Ministry undertakes various activities like prevention and control of pollution, conservation and survey of flora and fauna, forests and wildlife, protection of environment etc., in the framework of legislations. The Ministry works towards its desired objectives by conducting surveys, organizing regeneration programmes, collecting and disseminating environment information, creating awareness among individuals about pollution and its hazardous impacts.

The MoEF has constituted a number of pollution control acts for the prevention, control and abatement of different types of pollution in India. These acts are:

• The National Environment Tribunal Act, 1995 (27 of 1995).
• The National Environment Appellate Authority Act, 1997 (22 of 1997).
• The Water Prevention and Control of Pollution Act, 1974 (6 of 1974).
• The Water (Prevention and Control of Pollution) Cess Act, 1977 (36 of 1977).
• The Air (Prevention and Control of Pollution) Act, 1981 (14 of 1981).
• The Environment (Protection) Act, 1986 (29 of 1986).
• The Public Liability Insurance Act, 1991 (6 of 1991).

The MoEF is further divided into various divisions to achieve its objectives effectively. The different divisions related to the environment are as follows:

• Clean Technology
• Control of Pollution (CP)
• Environmental Education (EE)
• Environmental Impact Assessment (EEA)
• Environmental Information (EI)

  1. Environmental Information System (ENVIS) (http://punenvis.nic.in/)
  2. ENVIS—A gateway on Sustainable Development (http://punenvis.nic.in/index2.aspx?slid=56&mid=1&langid=1&sublinkid=35)
  3. Database of Environmental Experts in India 2007
  4. National Natural Resource Management System
  5. NGO Cell (NC)

• Environmental Research
• Policy and Law.

5.1.2 Clean Technology Division

In order to promote the development of clean technology, development of tools and techniques for pollution prevention and to formulate sustainable development strategies, the Ministry granted an aid in 1994 for the development and promotion of clean technologies. As against the conventional technologies, the cleaner technology aims at avoiding or minimizing the generation of pollution at the production process. They even make lesser use of the natural resources and eliminate emissions and waste.

The clean technology division has laid down several objectives for the adoption of clean technology in India. These include setting up more research and development institutes in India for the development, evaluation and adoption of these cleaner technologies, creating awareness about the existence of any such technology in India or abroad, providing the necessary financial support for the adoption of these technologies. The division has undertaken various projects under its stride since its inception in 1994. A few among these include Natural Resource Accounting Studies for Yamuna Sub-Basin by National Environmental Engineering Research Institute (NEERI); Life Cycle Assessment (LCA) Studies in Thermal Power Plants by Indian Institute of Environment Management, Navi Mumbai; and other pollution prevention and waste utilization strategies.

The MoEF does not provide any financial assistance to projects which involve primary research. However, financial assistance is provided to projects where primary research work has been completed and which are ready for pilot scale demonstrating research on any innovative technologies in the areas of highly polluting categories of industries. Furthermore, the MoEF has also formulated an evaluation and monitoring committee under the chairmanship of Professor L. Kannan, Vice Chancellor, Nagaland University for granting financial assistance to prospective proposals for the development and implementation of clean technology.

5.1.3 Control of Pollution Division

The pollution control division under the MoEF handles all matters connected with the prevention of pollution. It coordinates with the pollution boards of different states in India in ensuring that pollution levels in various states are below the prescribed limits. The main responsibilities of the Control of Pollution division include the following:

Administration of the various pollution control acts in India. These include the Water Act (1974), Air Act (1981) and the Environment Act (1986). These acts are discussed in details in the following paragraphs. The division also deals with litigations, court cases pertaining to matters on Air Act, Water Act, and Environment Protection Act.

- Dealing with all matters relating to the Central Pollution Control Board.
- Providing financial support to various state control boards in procuring scientific equipments to limit or prevent pollution. Financial assistance is also provided to the State Boards/state governments to deal with complaints on air, water and environment pollution. The division also analyzes the environment statement received from the state pollution control boards.

- Addressing complaints from people relating to any pollution issue.
- Monitoring and surveying the 17 highly polluting sectors and 22 critically polluted areas.
- Administering and dealing with financial matters relating to the National Environment Appellate Authority (NEAA).
- Formulating the noise pollution control standards.
- Matters on vehicular pollution emission standards. The division also formulates and reviews emission standards for various industrial units, automobiles etc., including water and air quality standards. The division lays emphasis on the adoption of clean technology in small scale industries.
- Ensuring adequate control on water pollution including marine pollution. It also deals with air and water quality monitoring and surveillance programme.
- Formulation of waste minimization programmes and environment management system. The division prepares the environmental action plan for specific areas.
- The division also works on the World Bank project which includes the schemes of Zoning Atlas, Air quality monitoring and pollution emission standards for industry.
- Dealing with all matters relating to the pollution of river which are not covered under the National River Conservation Programme (NRCP).
- Dealing with matters relating to environment health cell.

5.2 Various Laws to Control Pollution in India

Under Article 48A of the, 42nd Amendment Act under the Indian Constitution, the government of India provided for the protection of environment and forests. As per the Act, "The state shall endeavor to protect and improve the environment and to safeguard the forests and wildlife of the country". In addition to the above act, under the article 51A of the same amendment, under the fundamental duties of the citizens of India' the act states that 'it would be the fundamental duty of every citizen to protect and improve the natural environment including forests, lakes, rivers and wildlife and to have a natural compassion for living creatures'. The act came into force on 3 January 1977.

5.2.1 Water Act (Prevention and Control of Pollution Act, 1974)

The Water Act for the prevention and control of water pollution was the first regulation to be enacted in India with respect to pollution. The objective of the water act is to make provisions for the prevention and control of water pollution along with maintaining and restoring the wholesomeness of water. Furthermore it requires the establishment of Boards for the prevention and control of water pollution, for conferring on and assigning to such board powers and functions relating thereto and for matters concerned therewith.

Section (2) of the Water Act defines:

- Water pollution as the contamination of water or such alteration of the physical, chemical or biological properties of water or such discharge of any sewage or trade effluent or any other liquid, gaseous or solid substance into water (whether directly or indirectly) as may, or is likely to, create a nuisance or render such water harmful or injurious to public health or safety, or to domestic, commercial, industrial, agricultural or other legitimate uses, or to the life and health of animals or plants or of aquatic organisms.
- 'Sewage effluent' as effluent from any sewerage system or sewage disposal works and includes sullage from open drains.
- Trade effluent as any liquid, gaseous or solid substance which is discharged from any premises used for carrying on any industry, operation or process or treatment and disposal system.

The Central Board constituted by the Central Government under Section (3) of the Water Act shall have the requisite powers to perform the functions assigned to it under the Act.

The functions of the Central Board under Section (16) of the 'Water Act' comprise of the following:

- Advise the Central Government on any matter concerning the prevention and control of water pollution.
- Coordinate with the working of various State Boards by providing them with the technical assistance and guidance. Also to conduct sponsor investigations and research relating to the problems of water pollution and prevention, control or abatement of water pollution.
- Arrange for the training of persons engaged or to be engaged in programs for the prevention, control or abatement of water pollution.
- To regularly collect, compile and publish all relevant information and data relating to water pollution. Furthermore to work towards the technological advancements in the methods for effective prevention and control of water pollution.
- To lay down standards in consultation with the state governments for the quality of water, flow characteristics of the stream or well and the nature and use of the water in such stream or well or streams or wells.
- Organize nation-wide programmes for the prevention, control or abatement of water pollution.

The State Board as constituted by every state government in which the Water Act is implemented would have the requisite powers to perform the functions vested on it under the water Act of 1974.

The functions of the State Board under Section (17) of this Act consist of the following:

- To apprise the state government on any matter concerning the prevention, control and abatement of water pollution.
- To coordinate with the Central Board in organizing training of persons engaged or to be engaged in programmes relating to prevention, control and abatement of pollution.

- To conduct research and investigations relating to problems of water pollution, prevention, control or abatement of water pollution.
- To inspect sewage or trade effluents, works and plants for the treatment or sewage and trade effluents and to review plans, specifications, or other data relating to plants set up for the treatment of water, works for the purification thereof and the system for the disposal of sewage or trade effluents.
- To evolve economical and reliable methods of treatment of sewage and trade effluents, with due care of soils, climate and water resources of different regions.
- To explore ways and methods for utilization of sewage and trade effluents in agriculture.
- To lay down standards of treatment of sewage and trade effluents to be discharged into any particular stream taking into account the minimum fair weather dilution available in that stream and the tolerance limits of pollution permissible in the water stream, after the discharge of such effluents.
- To advise the State government with respect to the location of any industry, that is likely to pollute a water stream in that particular location.
- To perform other functions as may be prescribed by the Central Board and the State Government from time to time.

According to the provisions of Section (18) of this Act:

- The Central Board shall be bound by directions in writing given to it by the Central Government.
- Every State Board shall be bound by directions given to them by the Central or the State Governments.

Furthermore, where the Central Board is of the opinion that the State Board has defaulted in complying with any directions given to it by the Central Government and because of which an emergency has arisen then the Central Board may perform the functions of the State Board in relation to such area, such period and for such purposes.

As per Section (24) of this Act:

- No person shall knowingly cause or permit any poisonous, noxious or polluting matter determined in accordance with such standards as may be laid down by the State Board to enter into any stream or well or on sewer or on land.
- No person shall knowingly cause or permit to enter into any stream any other matter which may tend, either directly or in combination with similar matters, to impede the proper flow of the water of the stream in a manner leading or likely to lead to a substantial aggravation of pollution due to other causes or of its consequences.

Under Section (25) of this Act:

- No person should try to set up an industry, operation or process or any disposal system which is likely to discharge sewage or trade effluents into a stream or well or sewer or on land.

- Bring into use any new or altered outlets for the discharge of sewage.

Every State Board is also required to maintain a register containing particulars or conditions imposed under this section and the contents of the register that relates to any outlet, or to any effluent, from any land or premises shall be open to inspection at all reasonable hours by any person interested in, or affected by such outlet, land or premises.

As per Section (32) of this Act, if it appears to the State Board, that there is a presence of any poisonous, noxious or polluting matter in any stream or well or on land by reason of the discharge of such matter in such stream, well or on such land, and if the State Board is of the opinion that it is necessary to take immediate action, then it may carry out any of the following operations:

- Removing the matter from the stream or well or on land and disposing it in such a manner as the Board considers appropriate.
- Remedying or mitigating any pollution caused by its presence in the stream or well.
- Issuing orders restraining or prohibiting the persons concerned from discharging any poisonous, noxious or polluting matter or from making in sanitary use of the stream or well.

5.2.2 Air Prevention and Control of Pollution Act, 1981

The Air Act was legislated in India in the year 1981 to monitor the quality of air in India and to take measures for the control, prevention and abatement of air pollution. The 'Air Act' came into force on the 1st April 1988. As per Section (1) of the Act, the Act applies to whole of India.

Section (2) of the Act defines the following terms as:

- Air pollutant is defined as the presence of any solid, liquid or gaseous substance in such a concentration/proportion which may prove harmful to the health of human beings, animals and other living creatures and plants and environment.
- Air pollution is defined as the presence of any air in the atmosphere.
- 'Approved appliances' refers to the use of any equipment or gadget used for generating or consuming fume and which is approved by the State Board for the purpose of the Act.
- Control Equipment refers to any apparatus, device or equipment or system to control the quality and manner of emission of any air pollutant and includes any device used for securing the efficient operation of any industrial plant.

As per the Section (3) of the Air Act, the Central Pollution Control Board (CPCB) for the prevention and control of water pollution, constituted under section 3 of the Water Act shall also act as a Central Board for the prevention and control of air pollution in India. The CPCB would have all the necessary powers to ensure the prevention, control and abatement of air pollution (http://cpcb.nic.in/).

Under Section (4) of the Air Act, any state which has a State Board for the control and prevention of water pollution, under the section 4 of the Water act, shall also act as a State Board for the prevention and control of air pollution, under section (5) of the Air Act, with all the required powers to perform its functions. For those states which do not have a State Board for the prevention and control of water, but are still abiding by the Water Act of 1971, are notified to constitute a State Board for the prevention and control of Air pollution.

The Central Pollution Control Board has also to declare any air pollution control area under the Air Act of 1981. The CPCB has also to lay down standards for treatment of sewage and trade affluent and for emission from automobiles, industrial plants and any other polluting source. The CPCB has also to assess the quality of ambient water and air and inspect waste water installation, air pollution control equipment, industrial plants or manufacturing processes to evaluate their performance and to take steps for the prevention, control and abatement of pollution. For the successful implementation of the Air Act the Board would meet at least once in every 3 months to ensure that all rules in the Act are duly followed.

As per Section (16) of the 'Air Act' the Central Pollution Control Board is assigned the following functions:

- Advise the Central Government on any matter relating to the prevention, control and abatement of air pollution. The Board is responsible for holding nationwide programmes for the purpose of ensuring control, prevention and abatement of air pollution.
- Coordinate with different State Boards, provide technical assistance and guidance, and conduct the necessary investigations and research to ensure adequate measures are being taken for air pollution control and also to resolve any disputes that may arise within the State Boards.
- Organizing adequate training programmes for individuals who would engage in programmes for the control, prevention and abatement of air pollution.
- Organize nation-wide programmes for the prevention, control and abatement of air pollution.
- Lay down standards for ambient quality of air.
- Collect, compile and publish technical and statistical data relating to air pollution and to highlight measures for its effective prevention, control and abatement. Moreover the board has also to ensure that any information on pollution related matters like air pollution level alerts etc., are disseminated regularly to people through media or other means.
- The Central Board has to abide by any directions in writing given to it by the Central Government.

Section (17) of the 'Air Act' defines the functions of the State Boards towards controlling Air pollution as follows:

- Apprise the state governments on all matters relating to the prevention, control and abatement of air pollution. In addition the State Boards have also to advise the state governments on the feasibility of any location or premises from the emission of air pollutants point of view, for setting up an industry.

- Coordinate with the Central Board in disseminating pollution related information among masses. To organize training programmes in coordination with the Central Board for individuals to be involved in the control, abatement and prevention of air pollution programmes.
- Power to inspect any time, any industrial unit, manufacturing plant to ensure that the air quality standards are met and to take steps where ever necessary for the control, abatement and prevention of air pollution.
- Lay down standards for the emission of air pollutants into the atmosphere from industrial plants, automobiles or for the discharge of air pollutants from any other source.
- To ensure that all the functions are being carried out in a timely manner. Furthermore to ensure that any task towards air pollution control and abatement prescribed by the Central Board, state governments from time to time is carried out satisfactorily.
- To adhere by the directions in writing given to it by the state government or the Central Board. However if the State Board fails or defaults in complying with the directions given to it by the Central Board and an emergency situation has arisen because of it, then the Central Government can give orders to the Central Board to perform any of the functions of the State Board in relation to such area, for such period and for such purposes.

Under sub-section (1) of Section (19) of the 'Air Act', state governments have the power to declare any area within a state as pollution sensitive area, or air pollution control area after due consultation with the State Board. If the state government after due consultation with the State Board is of the opinion that any fuel, is likely to cause air pollution in any air pollution control area, it may by notification in the official gazette prohibit the use of such fuel in such area with effect from such date as prescribed in the notification. Similarly if the state government after consultation with the State Board is of the opinion that the burning of any material apart from fuel is likely to cause emission of air pollutants in the air pollution control area, then it may by notification in the official gazette prohibit the burning of such material in such area. Any disputes/inconsistencies between the Central and the state boards in the discharge of their functions would be taken care of by the Central government.

As per section (21) of the Act, no industrial unit can set up a plant in the air pollution control area without the prior consent of the State Board. Under Section (22 A) of the Act if the State Board finds that the emission of air pollutants is in excess of the standards laid down by the State Board, the State Board may make an application to the court restraining such person or industrial unit from emitting such air pollutants.

A State Board or any officer empowered by it in this behalf, under Section (26) of the Act, have the power at all times to take samples of air or emissions from any chimney, flue or duct or any other outlet for the purpose of analysis of the air pollutants discharged.

5.2.3 The Environment Protection Act, 1986

The Environment Protection Act for the protection and improvement of environment and for matters connected therewith was enacted in the year 1986. Under section (1) of the Act, it extends to the whole of India. This Act of Parliament got consent from the President of India on the 23rd May 1986.

Under the section (2) of the Act,

- Environmental pollutant is defined as the presence of any solid, liquid or gaseous substance present in such concentration as may be or tend to be injurious to environment.
- Environment pollution refers to the presence of any environmental pollutant in the atmosphere.

As per section (3) of the Act, all the necessary powers for the purpose of protecting and improving the quality of the environment and preventing, controlling and abating environment pollution are vested with the Central Government.

The following are considered to be the functions of the Central Government under the Section (3) of the Act:

- Coordinating with various state governments, officers and other authorities under this act, or the rules made there under.
- Organizing and planning nationwide programmes for the prevention, control and abatement of environmental pollution.
- Laying down standards for the quality of environment for the prevention, control and abatement of pollution. This includes laying down standards of emissions from different sources taking care of the quality or composition of the emission or discharge of environment pollutants from such sources.
- Providing clear guidelines on areas or regions where any industrial operations cannot be carried out and if industrial operations do take place then to ensure that adequate precautions are taken for the same.
- Laying down procedures and safeguards for the prevention of accidents which may cause environment pollution and mentioning the remedial measures for such accidents.
- Laying down procedures and safeguards for the handling of hazardous substances.

The Central Government under sub-section (3) of section 3 may appoint officers with such designations as it thinks fit for the purposes of this Act and may entrust to them such powers and functions under this Act as it may deem fit. The Central Government may, in the exercise of its powers under this Act, issue directions in writing to any person, officer or any other authority and such person, officer or authority shall bound to comply with such directions. The Environment protection Act does not require the institution of the Central Board for the same. Under Section (6) of this Act, the Central Government may make rules in respect for all or any of the following matters through notification in the Official Gazette:

- The air, soil and water quality standards for various areas and purposes.
- Maximum allowable limits of concentration of various environmental pollutants.
- The procedures and safeguards for handling of hazardous substances.
- Prohibition and restriction on the handling of hazardous substances.
- Prohibition and restrictions on the location of industries.
- Procedures and safeguards for the prevention of accidents which may cause environment pollution and providing remedial measures for such accidents.

Sections (7) and (8) of this Act require that:

- No person carrying any industry, operation or process shall discharge or emit or permitted to discharge any environment pollutant in excess of such standards as may be prescribed.
- No person shall handle or cause handling any hazardous substance except in accordance with such procedure and after complying with such safeguards as may be prescribed.

As per Section (9) of the Act in a situation where the discharge of any environment pollutant is in excess of the prescribed standards or is expected to occur due to any accident or other unforeseen act or event, the person responsible for such discharge and the person in charge of the place at which such discharge occurs or is apprehended to occur shall be bound to prevent or mitigate the environment pollution caused as a result of such discharge and shall also forthwith:

- Intimate the fact of such occurrences or apprehensions of such occurrence;
- Be bound if called upon, to render all assistance, to such authorities or agencies as may be prescribed.

On receiving such information with respect to the occurrence of any such environment pollution due to the discharge of any environment pollution in excess of the prescribed standards, either through intimation or otherwise, the authorities or agencies referred to in sub-section (1) shall, as early as practicable, because such remedial measures to be taken as are necessary to prevent or mitigate the environment pollution. The expenses incurred on any remedial measures taken by the authorities or agencies together with interest from the date when the demand for the expenses is made until it is paid may be recovered by such authority or agency from the person concerned as arrears of land revenue or of public demand.

As per Section (10) of the Act, any person empowered by the Central Government in this behalf shall have the right to enter any place for the purpose of examining and testing any equipment, industrial plant, record, register, document or any other material object or for conducting a search of any building in which he has reason to believe that any offence under this Act or the rules made there under has been or is being or is about to be committed and for seizing any such equipment, industrial plant, record, register, document that it may furnish evidence of the commission of an offence punishable under this Act or the rules made there under or that such seizure is necessary to prevent or mitigate environmental pollution. Moreover any person carrying on any industrial operation or handling any hazardous substance is bound to render all assistance to the person empowered by the Central Government. If the person fails to do so then the person shall be guilty of the offence under this Act.

Under section (11) of the Act, the Central Government or any of its officer empowered by it in this behalf, shall have the power to take samples of air, water, soil or any other substance from any of the factory, premises or any other place for the purpose of analysis. The person taking the sample shall specify to the person in charge of the place his intentions for taking the sample for analysis purposes.

According to Section (15) of the Act, any person whosoever if fails to comply with or contravenes any of the provisions of this Act, or the rules made or orders or directions issued there under, shall, in respect of each such failure or contravention, be punishable with an imprisonment of up to 5 years or a fine of up to one lakh Rupees or both. In case the failure, contravention continues, there would be an additional fine which may extend to five thousand Rupees for every day during which such failure or contravention continues after the conviction of the first such failure or contravention, with an imprisonment of up to 7 years in case the failure extends beyond 1 year.

Under Section (16) of the Act, if an offence under the Act is committed by a company, then every person in the company, who at the time of the offence was committed, was directly in charge of, and was responsible to the company for the conduct of the business of the company shall be deemed to be guilty of the offence and liable to be punished accordingly.

Under Section (17) of the Act, if any Department of the Government is responsible for committing offence under the Act, the Head of the Department shall be deemed guilty of the offence and shall be liable to be punished accordingly.

The State Government or any other authority or officer, under Section (20) of this Act, shall be liable to furnish any report, returns, statistics, accounts and other information to the Central Government as and when it requires.

The Central Government, under Section (25) of this Act, may by notification in the Official Gazette make rules on all or any of the following matter:

- The standards in excess of which the environmental pollutants shall not be discharged.
- The procedure and safeguards for handling hazardous substances.
- The authorities or agencies to which the knowledge of the occurrence or the likely occurrence of the discharge of any environment pollutant in excess of the prescribed standards shall be given. Moreover all assistance would also be rendered accordingly.
- The manner for taking samples of air, water and soil or other substance for the purpose of analysis shall be taken.

5.2.4 The Environment (Protection) Rules, 1986

These rules were formulated by the Central Government in exercise of the powers conferred by Sections 6 and 25 of the Environment (Protection) Act, 1986. Under

Section (3) of these rules, for the purpose of protecting and improving the quality of the environment and preventing and abating environment pollution, the standards for emission or discharge of environmental pollutants from the industries, operations or processes is specified.

5.2.5 The National Environment Tribunal Act, 1995

This Act was constituted in the year 1995 with the objective of providing strict liability arising out of any accident occurring in handling hazardous substances and for the establishment of a National Environment Tribunal for quick and effective disposal of cases arising from such accidents, with a view to give relief and compensation for damages to person, property and environment and for matters connected therewith or incidental thereto.

As per Section (2) of the Act,

- 'Accident' is defined as an accident involving a sudden or unexpected or unintended occurrence while handling any hazardous substance resulting in continuous or intermittent or repeated exposure to death of, or injury to, any person or damage to any property or environment.
- 'Hazardous Substance' means any substance or preparation which is defined as hazardous substance in the Environment (Protection) Act, 1986 and exceeding such quantity as specified by the Central Government under the Public Liability Insurance Act, 1991

Under Section (3) of the Act, if there is death or injury to any person or damage to any property or environment, from an accident' the owner shall be liable to pay compensation for such death, injury or damage. If the death, injury caused an accident is not due to individual activity but the combined or resultant effect of several such activities, operations and processes, the Tribunal be equitably divide the liability for compensation among those responsible for such activities. For any compensation awarded by the Tribunal on grounds of damage to the environment shall be remitted, as per Section (22) of the Act, to the authority specified under sub-section (3) of section 7A of the Public Liability Insurance Act, 1991 for being credited to the Environmental Relief Fund established under that section.

Under Section 3(1) of the Act, the compensation for damages may be claimed under any of the following:

- Death, permanent, temporary, total or partial disability or other injury or sickness.
- Loss of employment, business or both. Also loss of wages due to total or partial disability or permanent or temporary disability.
- Medical expenses incurred for treatment of injuries, sickness.
- Damage to private property.
- Expenses incurred by the government or any local authority in providing relief aid and rehabilitation to the affected persons.

- Expenses incurred by government for any administrative or legal action to cope with any harm or damage, including compensation for environmental degradation and restoration of the quality of the environment.
- Claims on account of any harm, damage or destruction to the fauna including milch and draught animals.
- Claims on account of any harm, damage or destruction to flora including aquatic flora, crops, vegetables, trees and orchards.
- Claims including cost of restoration on account of any harm or damage to environment including pollution of soil, air, water, land and ecosystems.
- Loss and destruction of any property other than private property.
- Any other claim arising out of, or connected with, any activity of handling hazardous substance.

The application for claim for compensation as per Section (4) of the Act can be made by any of the following:

- The person who has got the injury.
- Owner of the property to which damage is caused.
- In case of the death, by the legal representatives of the deceased, whether any person or the owner of a property.
- Any organization or body functioning in the field of environment and recognized in this behalf by the central government, or by the central or the state government itself.

5.2.6 The National Environment Appellate Authority Act, 1997

This Act was initiated in the year 1997, with the objective of establishing a National Environment Appellate Authority for hearing appeals with respect to restriction of areas in which any industries, operations or processes or class of industries, operations or processes shall not be carried out subject to certain safeguards under the Environment (Protection) Act, 1986 and for matters connected therewith or incident thereto. This Act came into force on the 30th of January 1997. As per Section (3) of the Act, the Central Government by notification in the official Gazette establishes the National Environment Authority to exercise the powers conferred upon it, and to perform the functions assigned to it under the Act.

As per Section (11) of the Act, any individual dissatisfied by an order granting environment clearance in the areas in which any industries, operations or processes or class of industries shall not be carried out or shall be carried out subject to certain safeguards, may appeal to the Authority within 30 days from the date of such order. Under Section (15) of the Act, no civil court or other authority shall have jurisdiction to deal with any appeal in respect of any matter which the 'National Environment Authority' is so empowered by this Act.

If any offence under this Act is committed by a company then, every person directly in charge of and responsible for the business of the company, at the time of the offence, shall be punishable according to Section (20) of the Act. Furthermore, if an offence is committed by a company and it is proved that the offence has been committed with the consent of any director, manger, secretary or any other officer of the company, shall also be deemed guilty of the offence and shall be liable to be punished accordingly.

5.2.7 The Noise Pollution (Regulation and Control) Rules, 2000

The Central Government in exercise of the powers conferred by Section 3, 6 and 25 of the Environment (Protection) Rules, 1986 and with rule 5 of the Environment (Protection) Rules, 1986 made the following rules for the regulation and control of noise producing and generating sources.

As per rule 3 of this Act, the ambient air quality standards in respect of noise for different areas/zones are specified below:

Area code	Category of area/zone	Limits in dB(A) leq*	
		Day time	Night time
(A)	Industrial area	75	70
(B)	Commercial area	65	55
(C)	Residential area	55	45
(D)	Silence zone	50	40

'Day Time' shall mean from 6.00 a.m to 10.00 p.m; 'Night Time' shall mean from 10:00 p.m to 6.00 a.m; * dB (A) Leq denotes the time weighted average of the level of sound in decibels on Scale A which is relatable to human being.; 'A' in dB (A) Leq, denotes the frequency weighting in the measurement of noise and corresponds to frequency response characteristics of the human ear.

Also Under rule 3 of this Act:

- The State Government has to take measures for the abatement of noise including noise emanating from vehicular movements, blowing of horns, bursting of crackers, use of loud speakers or public address system, and sound producing instruments and also to ensure that the existing noise levels do not exceed the ambient air quality standards specified above.
- Also a silence zone is defined as an area comprising not less than 100 metres around hospitals, educational institutions, courts, religious places or any other area which is declared as such by the competent authority.
- The noise level at the boundary of the public place, where loudspeaker or public address system or any other noise source is being used shall not exceed 10 dB (A) above the ambient noise standards for the area or 75 dB (A) whichever is lower.

Under the Noise Pollution (Regulation and Control) Amendment Rules, 2009, the State Government shall take measures to prevent the blowing of horn at night time in silence zones and residential areas except during an emergency. Under rule 4 of this Act, the authority shall be responsible for the enforcement of noise pollution control measures and for ensuring due compliance with the ambient air quality standards with respect of noise.

Furthermore as per the Noise Pollution (Regulation and Control) Amendment Rules, 2006, the respective State Pollution Control Boards in consultation with the Central Pollution Control Board shall collect, compile technical and statistical data relating to noise pollution and measures devised for its effective prevention, control and abatement, under rule 4 of this Act.

Under rule 5 of this Act, a loudspeaker or a public address system shall not be used except after obtaining written permission from the authority. Also a loud speaker or a public address system cannot be used at night (between 10 p.m and 6.00 a.m.) except in closed premises for communication within, e.g. auditoria, conference rooms, and community and banquet halls or during a public emergency. There would be no blowing of horns or bursting of crackers during night time in the silent zones/areas and residential areas except during public emergency.

As per rule (7) of this Act, any person can make a complaint to any officer authorized by the Central Government, or by the State Government in accordance with the laws in force and includes a District Magistrate, Police Commissioner, or any other officer designated for the maintenance of the ambient air quality standards, if the noise level exceeds the ambient noise standards by 10 dB (A) or more given in the corresponding columns above against any area/zone. The authority then shall act on the complaint and take action against the violator in accordance with the provisions of these rules and any other law in force.

Under rule 6(A) of this Act, whosoever violates any provision of these rules regarding restrictions imposed during night time shall be liable for penalty under the provisions of the 'Act'.

5.2.8 Biological Diversity Act, 2002

The Biological diversity act, with the objective of conservation of biological diversity, sustainable use of its components, and fair and equitable sharing of the benefits arising out of the use of biological resources, knowledge and matters connected therewith or incidental thereto was initiated in the year 2002. Under Section (1), this Act is valid for whole of India.

Under Section (2) of this Act:

- Biological diversity means the variability among living organisms from all sources and ecological complexes of which they are part and includes diversity within species or between species and of ecosystems.

- Biological resources means plants, animals and microorganisms or parts thereof, their genetic material and byproducts (excluding value added products) with actual or potential use or value but does not include human genetic material.

The National Biodiversity Authority established by the Central Government under Section (8) of this Act, may as per Section (18) of this Act can:

- Advise Central Government on matters relating to the conservation of biodiversity, sustainable use of its components and equitable sharing of benefits arising out of the utilization of biological resources.
- Advise the State Governments in the selection of areas of biodiversity importance.

Under Section (22) of this Act, the various state governments can establish their respective State Bio-diversity Boards. The state of Punjab has established, Punjab Biodiversity Board. Under Section (23) of this Act, the functions of the State Biodiversity Board would be:

- Advise the State governments, subject to any guidelines issued by the Central Government on matters relating to the conservation of biodiversity, sustainable use of its components and equitable sharing of the benefits arising out of the utilization of bio-logical resources.

Under Section (36) of this Act, the Central Government shall develop national strategies, plans and programmes for the conservation, promotion and sustainable use of the biological diversity including measures for identification and monitoring of areas rich in biological resources, incentives for training research and public education to create awareness with respect to biodiversity. Wherever the Central Government feels that the biological diversity or biological resources are being threatened by overuse, abuse or neglect, then it can issue directives to the concerned State Government to take immediate corrective measures along with any technical or other assistance which the State Government may need. The Central Government shall also undertake measures:

- To analyze the environmental impact of the project which is likely to have an adverse impact on the biological diversity, with a view to avoid or minimize such effects and wherever necessary provide for the public participation is such assessment.
- To regulate, manage and control the risk associated with the use and release of living modified organisms resulting from biotechnology, likely to have adverse impact on the conservation and sustainable use of the biological diversity and human health.

Any person, whosoever, if fails to abide by the directions and orders given by the Central Government, State Government, National Biodiversity Authority or the State Biodiversity Board shall under Section (56) of this Act, be punishable.

5.3 Central Pollution Control Board (CPCB)

The CPCB is the 'Central Board' for the prevention, control and abatement of air and water pollution in India. The Central Pollution Control Board (CPCB) was constituted in September 1974, under the Water Prevention and Control Act of 1974. The board was later also assigned the functions and powers under the Air Protection and Control Act of 1981.The primary function of the CPCB under the Water and Air Act is to emphasize and promote the prevention, control and abatement of water and air pollution respectively.

5.3.1 Functions of the Central Board

In addition to the main functions of promoting cleanliness of streams and wells and improving the quality of air and to prevent control or abate air pollution, CPCB has been assigned following functions:

- Advise the Central Government on any matter concerning prevention and control of water and air pollution and improvement of the quality of air;
- Plan and cause to be executed a nation-wide programme for the prevention, control or abatement of water and air pollution;
- Co-ordinate the activities of the State Boards and resolve disputes among them;
- Provide technical assistance and guidance to the State Boards, carry out and sponsor investigations and research relating to problems of water and air pollution, and for their prevention, control or abatement;
- Plan and organize training of persons engaged in programmes for prevention, control or abatement of water and air pollution;
- Organize through mass media, a comprehensive mass awareness programme on prevention, control or abatement of water and air pollution;
- Collect, compile and publish technical and statistical data relating to water and air pollution and the measures devised for their effective prevention, control or abatement;
- Prepare manuals, codes and guidelines relating to treatment and disposal of sewage and trade effluents as well as for stack gas cleaning devices, stacks and ducts;
- Disseminate information in respect of matters relating to water and air pollution and their prevention and control;
- Lay down, modify or terminate, in consultation with the state governments;
- Concerned, the standards for stream or well, and lay down standards for the quality of air;
- Establish or recognize laboratories to enable the Board to perform, and;
- Perform such other functions as and when prescribed by the Government of India.

For the successful discharge of its functions the CPCB formulated the National Ambient Air Monitoring Programme (NAMP). Under this programme the CPCB finds out about the air quality status and trends in different parts of the country and also takes measures to control the emission of pollutants from the industries and other sources and to keep them within the air quality standards. Furthermore the background air quality data is also provided to facilitate the setting up of industrial units and town planning. With regard to the Water quality standards, the CPCB initiated the Water Quality Monitoring and Surveillance Programme.

5.3.2   National Ambient Air Monitoring Programme (NAMP)

CPCB has initiated a nationwide programme of ambient air quality monitoring called NAMP. The objectives of this programme are:

- To determine the status and trends of ambient air quality,
- To determine whether the ambient air quality standards are violated,
- To identify non-attainment cities,
- To obtain knowledge and understanding for developing preventive and corrective measures and
- To understand the natural cleansing process undergoing in the environment through pollution dilution, dispersion, wind based movement, dry deposition, precipitation and chemical transformation of pollutants generated.

The programme covers three hundred and forty two operating stations covering one hundred and twenty seven cities in twenty six states and six Union Territories. Four air pollutants, namely Sulphur Dioxide (SO_2), Oxides of Nitrogen (NO_2), Suspended Particulate Matter (SPM) and Reparable Suspended Particulate Matter (RSPM/PM_{10}) have been identified for regular monitoring at all the locations. Meteorological Parameters like wind speed, wind direction, relative humidity and temperature were also monitored.

The monitoring of pollutants takes place for 24-h (4-h sampling for gaseous pollutants and 8-h sampling for particulate matter) with a frequency of twice a week to have one hundred and four observations in a year. The monitoring takes place with the help of the Central Pollution Control Board, State Pollution Control Boards, Pollution Control Committees and National Environmental Engineering Research Institute (NEERI), Nagpur. The CPCB coordinates as well as provides all the technical and financial support to these agencies for ensuring uniformity and consistency of the air quality of data monitored.

5.3.3   Water Quality Monitoring and Surveillance Programme

This programme consists of 1,019 stations in 27 states and six Union Territories. The monitoring is done on quarterly or monthly basis in surface waters and on half yearly basis in ground waters. 200 rivers, 60 lakes, 5 tanks, 3 ponds, 3 creeks, 13 canals, 17 drains and 321 wells are covered for monitoring under the programme.

At present as per the CPCB, the inland water quality monitoring network is operated under a three tier programme:

- Global Environment Monitoring System (GEMS)
- Monitoring of Indian National Aquatic Resources System (MINARS)
- Yamuna Action Plan (YAP).

5.4 Punjab Pollution Control Board (PPCB)

The Punjab Pollution Control Board (PPCB) was constituted in the year 1975, under Section 4 of the Water (Prevention and Control of Pollution) Act, 1974. The PPCB is the main governing body in Punjab for ensuring that the national ambient air quality standards are met. It works in close coordination with the Government of Punjab, in ensuring that any obstacles or hazards to clean air in Punjab are addressed in a timely fashion. During the Tenth Plan, government of Punjab provided a sum of Rs. 572 lakh to the PPCB towards its operations and Rs. 85 lakh in the Annual Plan 2004–2005 (http://www.punjabgovt.gov.in/).

The PPCB has three zonal offices and twelve regional offices. The PPCB has constituted the following cells for the effective implementation of the policies and decisions taken by the Board:

- Consent Management Cell
- Administrative Cell
- Finance and Accounts Cell
- Legal Cell
- Scientific Cell
- Hazardous Wastes Management Cell
- General Planning and Computer Cell
- Construction Cell
- Computer Section.

The Punjab Pollution Control Board abides by the following Acts for the control of environment pollution in the state of Punjab:

- *The Water (Prevention and Control of Pollution) Act, 1974 as amended till date.*
- *The Water (Prevention and Control of Pollution) Cess Act, 1977.*
- *The Air (Prevention* and Control of Pollution) Act, 1981 as amended till date.

In addition to the above Acts, the Ministry of Forests and Environment has also laid down the following rules for the management of hazardous wastes, Bio medical waste, solid waste management, recycled plastic, used batteries, noise pollution control and protection of the ozone layer under the environment. The objectives of the Punjab Pollution Control Board in pursuing its objective of abating and preventing pollution in Punjab are as follows:

- To control pollution at source with due regard to techno-economic feasibility for liquid effluents as well as gaseous emissions.

- To ensure that natural waters are not polluted by the discharge of untreated city sewage.
- To maximize the reuse of sewage and trade effluents and to use the treated effluent for irrigation and for industrial purposes.
- To minimize pollution control requirements through judicious location of new industries and relocation of industries wherever necessary.
- To control and minimize the pollution of air and water and to maintain the quality of air and water for designated use and purposes.

The strategy of the Punjab Pollution Control Board in controlling environment pollution in Punjab includes

- To deal with highly polluted areas of the state and highly polluted river stretches on priority basis for the control of pollution.
- To identify the various sources of pollution and to take measures for the abatement, control and prevention of pollution.
- To create awareness about environment pollution among local authorities, industries and people and to motivate them to take preventive measures for the control of pollution.
- To adopt measures for the control of pollution by adopting cost effective and less polluting technologies.
- To enhance the pollution control activities through training of manpower on pollution related matters and development of laboratories.

The functions of the Punjab Pollution Control Board in its pursuit of controlling and preventing pollution in Punjab include the following:

- To plan a comprehensive program for the abatement, control and prevention of pollution in Punjab and secure executions thereof.
- To apprise the industrialists and local authorities on information relating to pollution and assist them in adopting appropriate pollution control technologies and techniques.
- To create awareness among individuals about the benefits of clean and healthy environment and also to address public complaints on pollution.
- To support the development of pollution control technologies, eco friendly practices.
- To inspect sewage or trade effluent treatment and disposal facilities and air pollution control systems and to review plans, specifications or any other data relating to treatment plants, disposal systems and air pollution control systems in connection with the consent granted.

The Punjab Pollution Control Board has been monitoring the pollution levels at 20 locations out of which nine are in the residential cum commercial areas and 11 are in the industrial areas. As per the statistics of the period from 1995–2005, both the 24-h and annual averages of SPM/RSPM at residential cum commercial monitoring locations exceeded the permissible limits for residential areas (24 hourly permissible limits for SPM and RSPM are 200 and 100 μg/m^3 respectively and for Annual average permissible limits are 140 and 60 μg/m^3) throughout the year, with the maximum values being observed in Ludhiana followed by Mandi Gobindgarh, Jalandhar and Amritsar.

The Punjab Pollution Control Board has laid down guidelines with regard to pollution control for any entrepreneur wanting to set up an industrial unit in the state of Punjab. As per the provisions of the Water (Protection and Conservation) Act, 1974 and Air (Prevention and Control of Pollution) Act, 1981, any entrepreneur wanting to set up a new industrial unit or wanting to expand its existing industrial unit in the state of Punjabis required to obtain a 'consent to establish' (No Objection Certificate) from the Punjab Pollution Control Board. The Ministry of Forest and Environment has divided the industries in three different categories as per the pollutants being emitted by them. The three categories are: (i) Green Category (ii) Orange Category (iii) Red Category.

Industries falling in each of the three categories are mentioned in the **Appendix** Table. Industries which do not fall in any of the above three categories, the decision with regard to their categorization would be taken by the Punjab Pollution Control Board (PPCB).

The Punjab Pollution Control Board has divided the small scale industries into two categories namely, green and red categories, taking into account their potential pollution loads for determining the standards for establishing or expanding an industry. The procedure for obtaining consent for large or medium industry is same as that for a small industry falling in the red category. However in the case of the small scale red category industries, the decision to grant consent to establish or expand an industrial unit are taken at the zonal office level by the concerned senior environment engineer. However, for industries like brick kiln, dry rice sheller, cupola furnaces heat treatment units, the decision is made by the concerned regional office. With regard to the large and medium industries, the decision is made by the head office.

Moreover all industries whether large/medium or small are as per the Factories Act, 1948 require to obtain site clearance from the site appraisal committee (SAC) before obtaining the consent from the Punjab Pollution Control Board (PPCB). Furthermore, any new entrepreneur wishing to establish a new project or expand an existing one shall also check whether his category of industry falls under Schedule 1 appended to the EIA notification No. SO (60)-E dt. 27.1.1994 as amended on 4.5.1994, specified in the **Appendix** Table. If so, the entrepreneur is required to follow the procedure of Environment clearance also. In case the industry is among the one mentioned in **schedule 1 of the Appendix** Table, then the entrepreneur is required to obtain environment clearance from the Ministry of Environment and Forests. The application shall be made in the specified Performa along with the project report which should include the Environment management plans.

5.5 Punjab State Council for Science and Technology

The Punjab State Council for Science and Technology was established on 21 July 1983, with the objective of infusing scientific knowledge in the minds of people. The institute has been trying to achieve this through various means of display and publications, about the nature of life while signifying the useable aspects of available technologies.

Some of the main objectives of the institute include the following:

- Conservation of environment
- Pollution Control in the state of Punjab
- Providing consultation to various industrial units for undertaking development.

The institute tries to achieve these objectives by working towards the development of new technologies, providing technical support to the state government on development through development of science and technology etc. The institute focuses in providing both formal and informal assistance to the industrial and agricultural sector in carrying out their activities, in such a manner to ensure judicious utilization of natural resources with the least stress on the environment. The institute is trying to resolve the problems of water logging, chemically over saturated soils and their deteriorating fertility, stagnating agricultural productivity, ground water depletion and its pollution, selenium toxicity, conservation of eco-systems etc., persistent in Punjab as early as possible. Also, in future the institute aims at focusing on matters such as pollution control in Punjab, biotechnology, nanotechnology, and socio-economic development. The institute is divided into five divisions in working towards its goals:

- Environment
- Biotechnology
- Popularization of Science
- Consultancy Cell
- Water Regime Management.

5.6 Environment Division

The division of environment assists the State Department of Environment, Government of Punjab in technical matters pertaining to environment, identification of major areas of ecological concern, defining the state government policies and plans on various environmental issues, coordinating and monitoring schemes related to environment, creating environmental awareness and promoting environmental education, training and research. It is also implementing projects and programmes related to environment for international bodies like, UNESCO, UNDP, etc., as well as, programmes of the Ministry of Environment and Forests at the national level. A large number of projects are being undertaken by the institute under the Environment division (http://www.punjabgovt.gov.in/jsp/apps/work/Map pingOfMinistersPunjab.pdf).

Moreover Punjab State Council for Science and Technology was also recognized as one of the institutes for imparting training on pollution control, waste management, clean technologies, environment policies, health monitoring-and-assessment and solid waste management conducted by the Central Pollution Control Board under the human resource development programme. The first such training was conducted in December 2005 in Chandigarh. Since then this training exercise is expected to be a regular feature.

The Environmental Management Capacity Building-Environmental Information System (EMCB-ENVIS) node on State Environment issues was established at the Punjab State Council for Science and Technology in December 2002 under the World Bank assisted project (EMCB-ENVIS) of the Ministry of Environment and Forests for identifying the state of the environment and related issues. In January 2005, the node got upgraded to ENVIS Centre, under the sponsorship of the Ministry of Environment and Forest under the tenth 5 year plan.

5.7 Punjab Energy Development Agency (PEDA)

The Punjab Energy Development Agency was established in the year 1991, for the promotion and development of non-conventional and renewable energy programs or projects in the state of Punjab.

The objectives of PEDA in this regard are as follows (http://www.peda.gov.in/):

- Promotion, development and implementation of non-conventional energy technologies programs and projects.
- Promotion and development of Biomass/Agro residue based power projects.
- Implementation of a comprehensive energy conservation program in the industrial, agricultural, commercial as well as household sector.
- Promotion and implementation of new technologies for energy saving.
- Collection of energy data base to provide policy and planning input to the state government.
- Measures for improving the combustion efficiency of rice husk fired boilers.
- Analyze the availability and utility of biomass as energy source.
- Installing community/institutional biogas plants.
- Implementation of Integrated Rural Energy Program (IREP).

The projects undertaken by PEDA to meet its above objectives include the following:

- Mini hydel power generation.
- Solar energy based power generation projects.
- Biomass, Agro based power generation projects.
- Power generation from urban, industrial waste.
- Promotion and development of co-generation.
- Integrated rural energy program (IREP).
- Community institutional/Night soil biogas plants.
- National project on biogas development program.
- Solar Photovoltaic (SPV) water pumping systems.
- Solar cooker implementation program.
- Biomass gasification program.

- National program on improved chulah.
- Energy conservation study/feasibility study/Energy audit in the industry and other user sectors.
- Solar passive architecture–PEDA office complex.
- Power generation potential from non-conventional energy sources.

5.8   Punjab Biodiversity Board

The Punjab Biodiversity Board was notified in the state in December 2004 under section 22 of the Biological Diversity Act, 2002, to protect Punjab's natural eco-systems and its flora and fauna. The Board has been set up in the Department of Environment to ensure that biodiversity in both wild and cultivated areas are properly protected. Under the Act, no corporate body or association can commercially utilize the state's biodiversity without approval of State Biodiversity Board. Further, no foreigner without the approval of the National Biodiversity Authority (NBA) can obtain any biological sample or knowledge associated for research or for commercial utilisation or for bio-survey and bio-utilisation. These include wild relatives of crop species also. The Board has already notified committees to identify biological heritage sites outside Protected Area Network (PAN) and for identifying commercially important flora and fauna in the state.

Some of the functions of the board in meeting its objective are:

- To promote biodiversity conservation activities in both agriculture and wild areas.
- To implement the provisions of the Biological Diversity Act, 2002 in Punjab.
- To assist setting up of Biological Diversity committee at village and town level and expert committee at the state and district level.

The Punjab Biodiversity Board also maintains a database on the State's Biodiversity Strategy and Action Plan, Punjab's Environment status which includes both wild and agriculture biodiversity in the state of Punjab.

5.9   Summary of the Chapter

India is a legislation rich country with reference to pollution. The Ministry of Forest and Environment is a nodal agency in the administrative setup of the Union Government. The Ministry is entrusted with the task of planning, coordinating, overseeing and implementing various forestry and environment programmes. In order to promote the development of clean technology, development of tools and techniques for pollution prevention and to formulate sustainable development strategies, the Ministry granted an aid in 1994 for the development and promotion of clean technologies. Eleven major laws exist to control pollution in India and many forums for their implementation in various ways. Among the existing legislation on air pollution in India includes: Air Prevention and Control of Pollution Act, 1981; The Environment Protection Act, 1986; The National Environment Tribunal Act, 1995; The National Environment Appellate Authority Act, 1997; and Biological Diversity Act, 2002. Under these different Acts, provisions are made to protect the environment from all kinds of pollution related to industrial and agricultural activities. The Punjab Pollution Control Board (PPCB) is entrusted with the functions of planning a comprehensive program for the prevention, control and abatement of pollution in Punjab. PPCB has to support and encourage developments in the field of pollution control. PPCB has taken various measures to limit the amount of industrial pollution in the state but not much has been done to address agricultural pollution.

Chapter 6
Policies for Restricting the Agriculture Residue Burning in Punjab

Abstract This chapter highlights policies of the Punjab government to address crop stubble burning. Various departments like Punjab Agricultural University, Punjab Farmers Commission etc., are all making efforts to devise some alternate economic uses of rice stubble. Punjab government is also providing subsidy to the farmers to promote the use of equipments which help in checking the burning of crop residues. Similarly, Punjab Energy Development Agency is promoting non-conventional and renewable energy projects in the state that use crop waste as raw material.

Keywords SPM/RSPM levels · Policies to control air pollution · Alternate uses of rice stubble

In Punjab, industrial pollution, agricultural pollution and vehicular pollution are recognized as the three major contributors to air pollution. The air quality in Punjab is believed to be affected by industrial growth, urban growth and agricultural practices. The Punjab Pollution Control Board (PPCB) (http://www.ppcb.gov.in/index.aspx) monitors the pollution levels at 20 different locations in Punjab; nine of these locations are in the residential areas and eleven in the industrial areas. In this study we mainly focus on the air pollution generated in the residential cum commercial areas.

In this chapter we try to analyze the contribution of crop stubble burning in the emission of harmful gases and particulate matter into the air. Based on the findings, the policies of the Punjab government to address this are highlighted. With the looming problem of crop stubble burning, there is an urgent need to refer to the existing policies of the Punjab Government in place to address agricultural pollution. There is also a dire need to analyze the effectiveness of these polices in preventing farmers from burning their crop residues. Furthermore if the existing policies are found to be ineffective, what all policy measures can be suggested to the Government to put an end to this evil practice?

© The Author(s) 2015
P. Kumar et al., *Socioeconomic and Environmental Implications of Agricultural Residue Burning*, SpringerBriefs in Environmental Science,
DOI 10.1007/978-81-322-2014-5_6

The problem of crop stubble burning in Punjab is emerging as a major threat to not just the quality of air but also to the health of individuals in the state. The burning of crop stubble results in the emission of various harmful gases and particulate matter in the air. Burning of rice and wheat residue results in the emission of Suspended Particulate Matter (SPM), SO_2, NO_X and other harmful gases like Carbon monoxide (CO), CH_4 etc. As per an estimate by Gupta et al. (2004), one tonne of straw on burning releases 3 kg particulate matter, 60 kg of CO, 1,460 kg of CO_2, 199 kg of ash and 2 kg of SO_2. The objective of this chapter is to highlight the SPM, SO_2, NO_2 levels in Punjab with the National Ambient Air Quality Standards based on the existing data, projects undertaken by the Punjab Pollution Control Board. The study progresses further by looking into the existing policies of the Punjab government towards this and what suggestive measures can be made to abate this problem.

6.1 Monitoring and Recording the Levels of Pollution in Punjab

In Punjab at present as per the National Ambient Air Quality Monitoring program, three major pollutants are being monitored; these are the Suspended Particulate Matter (SPM) Respirable Suspended Particulate Matter (RSPM), Nitrogen Oxides (NO_2) and Sulphur Dioxide (SO_2). Other pollutants like Carbon Monoxide (CO), Ozone (O_3), Lead (Pb) and Green house gases like CO_2, CH_4 etc. are monitored depending on the availability of data. To assess the cumulative and overall impact of the three pollutants (SO_2, NO_2 and SPM) on air quality and also to assess the non cumulative non compliance of the standards, an Air Quality Index has been formulated. This index is measured as the sum of the ratios of the three major pollutant concentrations to their respective air quality standards (Table 6.1).

$$AQI = 1/3 * \frac{(SO_2)}{(SSO_2)} + \frac{(NO_2)}{(SNO_2)} + \frac{(SPM)}{(SSPM)}$$

The pollution levels in Punjab are also measured on the basis of the Exceedence Indicator. The Exceedence Indicator compares the pollutant concentrations in

Table 6.1 The rating scale for air quality index

Index value	Remarks
0–25	Clean air
26–50	Light air pollution
51–75	Moderate air pollution
76–100	Heavy air pollution
>100	Severe air pollution

Source Environment Indicators for Punjab (http://www.pscst.gov.in/)

different cities/towns with respective NAAQS and characterizes them into four broad categories based on the exceedence factor (Table 6.2).

As per guidelines by the Central Pollution Control Board (CPCB), the maximum concentration limit of SPM and RSPM in the residential, rural and other areas are 140 and 60 $\mu g/m^3$ respectively. However in Punjab SPM/RSPM are estimated as given in Fig. 6.1. The SPM level in Punjab at different residential cum commercial areas have always been above the maximum RSPM limits of 140 $\mu g/m^3$ for the years of study.

The SO_2 levels in Punjab at residential-cum-commercial areas have been below the maximum permissible limit set by the Central Pollution Control Board of 60 $\mu g/m3$ in residential/rural and other areas in all the years of study (Fig. 6.2). Similarly the NO_2 levels from 1997 to 2007 have been below the maximum permissible limit of 60 $\mu g/m3$ in residential/rural and other areas (Fig. 6.3).

The Central Pollution Control Board has set up National Ambient Air Quality Standards for all the states in India to follow. As per the Central Pollution Control Board (CPCB), all states in India have to abide by the National Ambient Air Quality Standards (Table 6.3). The National Ambient Air Quality Standards say that the levels of air quality in any region should be such so as to protect the public health, vegetation and property. Whenever and wherever two consecutive values exceed the limit specified above for the respective category, it would be

Table 6.2 The descriptive categories for different exceedence indicator values

Exceedence factor E.F	Remarks
>1.5	Critical pollution (C)
1.0–1.5	High pollution (H)
0.5–1.0	Moderate pollution (M)
<0.5	Low pollution (L)

Source Central Pollution Control Board, New Delhi (http://cpcb.nic.in/)

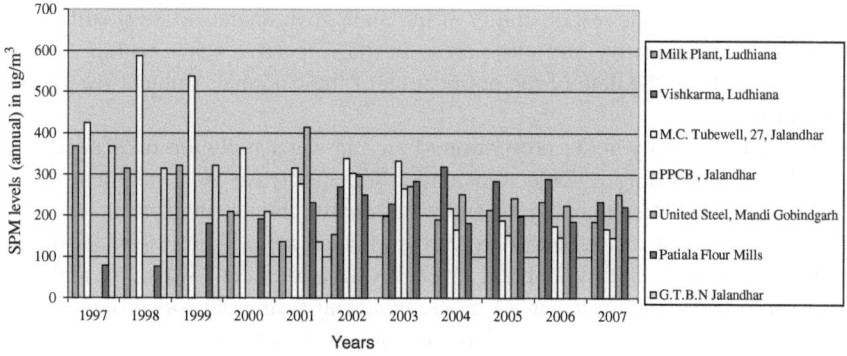

Fig. 6.1 SPM/RSPM levels at different residential cum commercial locations in Punjab.

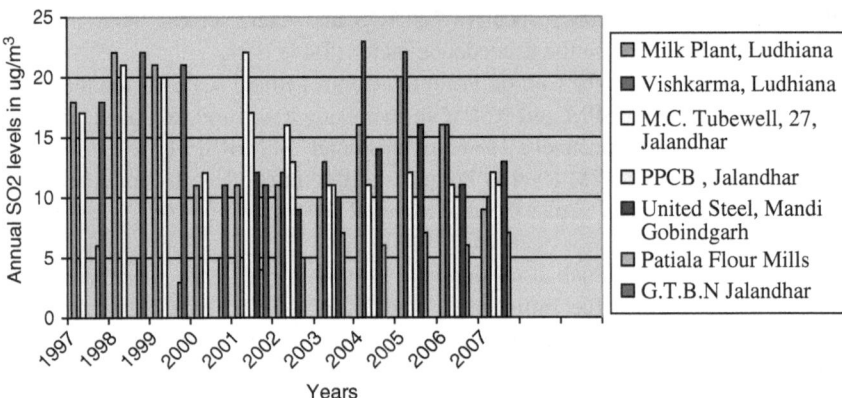

Fig. 6.2 SO$_2$ levels at different residential cum commercial areas in Punjab. Status of Environment and Related Issues

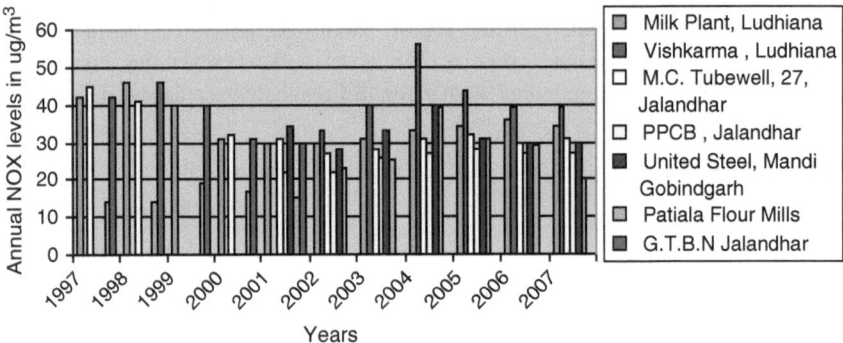

Fig. 6.3 NO$_2$ levels at different residential cum commercial areas in Punjab.

considered adequate reason to institute regular/continuous monitoring and further investigations. It is the responsibility of the State governments/State Board to intimate about the sensitive and other areas in the respective states within a period of 6 months from the date of the notification of the National Ambient Air Quality Standards.

The major activity under prevention of air and water pollution relates to grants to CPCB for fulfilling its objectives under Water Act, Air Act and Environment Pollution Act for strengthening air and water quality monitoring system. In 2008–2009, a total grant of Rs. 34.50 crore was planned for addition of water quality monitoring at 50 locations, ambient air quality monitoring at 75 stations, and automatic air quality monitoring in 11 cities. Financial assistance was provided to 15 SPCB/UTPCs for strengthening laboratory/other technical infrastructure under the scheme of abatement of pollution. To check environmental loss due to effluent

Table 6.3 National ambient air quality standards set by the CPCB

Pollutants	Time weighted average	Concentration in ambient air			Method of measurement
		Industrial areas	Residential, rural and other areas	Sensitive areas	
Sulphur dioxide (SO₂)	Annual average	80 µg/m³	60 µg/m³	15 µg/m³	Improved West and Gaeke method ultra-violet fluorescence
	24 h	120 µg/m³	80 µg/m³	30 µg/m³	
Oxides of nitrogen (NO₂)	Annual average	80 µg/m³	60 µg/m³	15 µg/m³	Jacob and Hochheiser modi-fied (Na-Arsenite) method
	24 h	120 µg/m³	80 µg/m³	30 µg/m³	Gas phase chemiluminescence
Suspended particulate matter (SPM)	Annual average	360 µg/m³	140 µg/nm³	70 µg/m³	High volume sam-pling (average flow rate not less than 1.1 m³/min)
	24 h	500 µg/m³	200 µg/m³	100 µg/m³	
Respirable particulate matter (RPM)	Annual average	120 µg/m³	60 µg/m³	50 µg/m³	Respirable particu-late matter sampler
	24 h	150 µg/m³	100 µg/m³	75 µg/m³	
Lead (Pb)	Annual average	1.0 µg/m³	1.00 µg/m³	0.75 µg/m³	AAS method after sampling using EPM 2000 or equivalent filter paper
	24 h	1.5 µg/m³	1.0 µg/m³	0.75 µg/m³	
Ammonia 1	Annual average	0.1 mg/m³	0.1 mg/m³	0.1 mg/m³	
	24 h	0.4 mg/m³	0.4 mg/m³	0.4 mg/m³	
Carbon monoxide (CO)	8 h	5.0 mg/m³	2.0 mg/m³	1.0 mg/m³	Non dispersive infrared (NDIR)
	1 h	10.0 mg/m³	4.0 mg/m³		Spectroscopy

Source Punjab Envis Center (http://punenvis.nic.in/)
Note Annual Arithmetic mean of minimum 104 measurements in a year taken twice a week 24 hourly at uniform interval

discharge, financial assistance was provided to 10 CETP to treat effluent emanat-ing from the clusters of compatible small scale industries.

Furthermore the Ministry also in its 2008–2009 Budget provided outlay of Rs. 3.05 crores, for the development and promotion of clean technology. The total outlay of the Ministry in 2008–2009 against total pollution control and preven-tion was Rs. 95.17 crores and Rs. 389 crores of plan and non plan budget, respec-tively. In addition, another Rs. 5 crores of non plan Budget was sanctioned by the

Ministry for supporting State Pollution Control Boards and State Environmental Departments for infrastructure development etc.

6.2 Existing Policies to Control Air Pollution

The Punjab Pollution Control Board (PPCB), Punjab State Council for Science and Technology (PSCST), Punjab Energy Development Agency (PEDA) (http://www.peda.gov.in/) are the Institutions that have been vested with the task of controlling pollution in Punjab. It is mainly the Punjab Pollution Control Board in coordination with the Central Pollution Control Board that advices the government on pollution related matters. During the 10th Five Year Plan Government of Punjab provided a sum of Rs. 572 lakh to the PPCB and Rs. 85 lakh for Annual Plan 2004–2005. The PPCB is the main governing body in Punjab for ensuring that the ambient air quality standards are met (http://www.punjabgovt.gov.in/).

6.2.1 Punjab Pollution Control Board (PPCB)

The main objectives of the Punjab Pollution Control Board in controlling pollution include effective control of air and water pollution, controlling pollution at source and to ensure that the pollution control standards are met. The PPCB is entrusted with the functions of planning a comprehensive program for the prevention, control and abatement of pollution. PPCB has to support and encourage developments in the field of pollution control. It has to make sure that the information collected by means of reports, projects on the pollution levels in various sectors is disseminated timely so that the Punjab government can formulate policies accordingly. Furthermore it should also encourage the development of machineries for ensuring pollution control in Punjab.

The Punjab Pollution Control Board has been monitoring the pollution levels at 20 locations out of which nine are in the residential-cum-commercial areas and 11 are in the industrial areas. As per the statistics of the period from 1995 to 2005, both the 24 h and annual averages of SPM/RSPM at residential-cum-commercial monitoring locations exceeded the permissible limits for residential areas (24 hourly permissible limits for SPM and RSPM are 200 and 100 $\mu g/m^3$ respectively and for Annual average permissible limits are 140 and 60 $\mu g/m^3$) throughout the year, with the maximum values being observed in Ludhiana followed by Mandi Gobindgarh, Jalandhar and Amritsar.

The Punjab Pollution Control Board abides by the Air Act of 1981 and the Environment Act of 1986 to control pollution in the state of Punjab. The Air Act has been adopted by the Government of Punjab to control environmental pollution in Punjab. The Punjab Pollution Control Board is entrusted with the task of ensuring that the above air laws are being followed in Punjab (http://www.punjabgovt.gov. in/jsp/apps/work/MappingOfMinistersPunjab.pdf).

6.2.2 Agriculture Councils

With the realization of the harmful effects of wheat and paddy monoculture on the ecology and environment, the Punjab government has been encouraging the diversification in agriculture, away from the rice-wheat cropping pattern towards other remunerative and less water intensive crops since the past few years. Basmati paddy, hyola, sunflower, pulses and vegetables are being promoted as alternative to paddy and wheat monoculture.

During the year 2005–2006, the Punjab state government in the wake of degradation of soil health, depletion of water reserves, faster erosion of the micronutrients reserves, caused due to the paddy wheat crop rotation, created the Agricultural Diversification Fund. Furthermore an outlay of 50.56 crores was provided in the Annual Plan of 2006–2007 to strengthen the agricultural infrastructure and speed up the process of agricultural diversification in the state. In the Annual Plan 2006–2007, a new programme, 'Agriculture Production Pattern Adjustment Programme in Punjab for Productivity and Growth' under the 12th Finance Commission was included with a budgetary provision of Rs. 24 crores per annum for four years till 2010.

During 2005–2006, the Punjab government devoted Rs. 10 crores for the creation of an 'Agricultural Research and Development Fund'. The funds were used for the development of better quality of alternative agricultural crops, improved agricultural practices and improved post harvesting management practices. In order to encourage farmers to use the crop stubble as fodder for animals and to meet the fodder requirements during the scarcity period an outlay of Rs. 20 lakh was sanctioned for the scheme enrichment of straw and cellulose waste.

In addition the Punjab Government in order to intensify its diversification programme in the agriculture sector has set up four Special Purpose Vehicles (SPVs) to promote citrus and fruit juices, value- added horticulture, viticulture and organic farming. These are:

- Council for Citrus and Agri. Juicing
- Council for Value added Horticulture
- Organic Farming Council of Punjab
- Viticulture Council of Punjab

These Councils were set up in January 2006 under Financial Commissioner (Development). The main objective of these Councils is to take measures for shifting Punjab from primary agricultural and low value produce to high value processed products. The Government's aim is that one-third of the State's farm sector should diversify to citrus and high value horticulture, viticulture and organic farming in the next 10 years.

6.2.3 Punjab State Council for Science and Technology

The Punjab State Council for Science and Technology (PSCST) caters to the science and technology requirements of the state. As per a report by the Council on the State Environment of Punjab 2006, the burning of crop stubble is banned in the state. In order to address the problem of agricultural waste, PSCST constituted a Task Force under the chairmanship of its Executive Director in September 2006. As per the Task Force, there is a need to adopt new ways and methods for better utilization of agriculture waste, especially rice stubble to mitigate the problem of the pollution caused due to burning of these residues in the fields. These include: strengthening of crop diversification program, alternate uses of agriculture residues through incorporation of paddy straw in soil by promoting Happy Seeder Technology, zero till etc., use of agricultural residues for power generation, use of paddy straw as protein enriched fodder for livestock after fermentation and biomethanization of paddy straw. The following actions have been initiated by various departments/institutions in the State in response to the recommendations of the Task Force:

6.2.4 Department of Agriculture

During the year 2008–2009, a total number of 5,117 farmers training camps at district, block and village level were organized by department of agriculture to make the farmers aware about the benefits of reincorporation of the crop residues. Besides this, the department was organizing frontline demonstrations to encourage farmers to adopt zero-till-drills, Happy Seeders, Rotavators and distribution of new agricultural implements on subsidy. During 2007–2008 and 2008–2009, 760 and 1,290 frontline demonstrations were organized at farmers' fields, respectively. As a result of these demonstrations and supply of these equipments on subsidy, these techniques were adopted by the farmers of the state in 5.92 and 7.21 lakh hectares area, respectively during the sowing of wheat crop in 2007–2008 and 2008–2009. To promote the use of equipments which help in checking the burning of crop residues, Rotavators, Happy Seeders, Zero-till-drills and Straw Reapers were distributed to the farmers on subsidy. During the year 2007–2008 in all, 2,659 rotavators, 1,383 Zero-till-drills, 2 Happy Seeders and 448 straw Reapers were distributed to the farmers on subsidy.

Further, the Department is promoting diversification of cropping pattern in Punjab under which area under Basmati rice had been increased from 1.5 lakh hectares to 3.5 lakh hectares in the past 5 years whereby straw of basmati rice can be used as a fodder.

6.2.5 Punjab Energy Development Agency (PEDA)

PEDA has been facilitating the setting up of 29 power projects with total installed capacity of 330 MW on BOO basis to private developers. These projects are being set up by the private developers with state-of-art technologies such as

Biomethanation, Combustion etc. The plants are designed to receive mixed waste such as paddy straw, cotton stalks and other agro residues available in the state. Out of these, one project of 8 MW had been commissioned in March 2009 and another of 14.5 MW in September 2009.

6.2.6 Department of Animal Husbandry

A total number of 2,478 demonstrations for protein enrichment of paddy/wheat straw through urea treatment had been conducted up to October 2008. Awareness camps for providing information about straw as animal bedding was also organized by the Department.

6.2.7 Punjab Agricultural University

PAU had been according priority towards developing efficient agro-technologies for crop residue recycling in machine harvested areas as an alternative to burning. The major equipments developed by PAU are: (i) Happy Seeder Machine for planting in standing paddy stubbles; (ii) Tractor Operated Paddy Straw Chopper; (iii) Straw Collector and Baler; (iv) Residue Incorporation in Soil; (v) Compositing Techniques using Paddy Straw.

(i) Happy Seeder Machine for planting in standing paddy stubbles: This technology, developed by PAU, has already been adopted by the government of Punjab and is being popularized by Department of Agriculture. Wheat was successfully sown in 200 acres area using Happy Seeder during 2007–2008 producing 5–10 % more yield (with 50–60 % less operational costs) compared to conventionally sown wheat. Financial analysis by PAU indicated that this machine is more profitable than other conventional alternatives like full stubble incorporation through direct drilling or rotary seeding.

(ii) Tractor Operated Paddy Straw Chopper: For incorporation of paddy straw into soil, the University has also developed a Tractor Operated straw Chopping-cum-spreading machine. The Machine, in a single operation, harvests the left over paddy stubble after combining, chops it into pieces and spreads it on to the field. The chopped and spread stubble then can easily be incorporated in the soil after light irrigation by using a rotavator or disc harrow and is allowed to decay.

(iii) Straw Collector and Baler: Baler is also another promising technology developed by PAU for collecting paddy straw. Balers make rectangular or round bales by collecting the loose straw from the ground. One operation of stubble shaver in a combine harvested paddy field, created favorable conditions for operating a baler, which in turn, results in smooth sowing of the next crop.

(iv) Residue incorporation in Soil: In situ incorporation of paddy straw before sowing wheat, did not adversely affect the wheat crop. Rather the incorporation of the residues had a favorable effect on soil physical, chemical and biological properties.

(v) Composting techniques using paddy straw: PAU has also been working on use of paddy straw as bedding material for animals and thereafter going in for its composting. A special machine has been acquired for the turnover of composting materials and its watering for a rapid generation of high value compost.

6.2.8 Punjab State Farmers' Commission

The Commission had approved the steps proposed by the Task Force. In order to reduce area under paddy without decreasing the income of the farmers, the commission has initiated the following programmes:

- Commercial dairy farming and increasing the area under fodder
- Production of vegetables under net house technology
- Encouraging cultivation of hybrid maize in kharif season
- Introduction of new high value crops such as banana cultivation in the state.

6.2.9 Department of Rural Development and Panchayats

The Department is popularizing technologies proposed by Department of Agriculture and PEDA and is facilitating provision of panchayati land for setting up of Biomass based power plants in the state. It had already facilitated 33 years lease of panchayati land of five villages for setting up such units.

Furthermore the Punjab State Council for Science and Technology has also set up a Consultancy Cell to address the problem of Pollution in Punjab. However this mainly caters to the air pollution caused by the industries. There have been various technological responses undertaken for conservation of environment affected by agriculture. These include the following:

6.2.10 Agriculture Diversification

It has been increasingly felt by the Government of Punjab, to move the farmers away from the rice-wheat crop rotation into new areas like vegetables, fruits, oil seeds, pulses, etc. The importance of crop diversification to protect the natural resources and to stabilize farm income is increasingly felt. The Government of Punjab in 2002 launched a multi crop multi-year contract farming scheme to give boost to crop diversification. The Punjab Agro Food Corporation (PAFC) has been implementing the task and it is believed that more than 0.186 million ha is covered under crops other than wheat and paddy like hoyla, winter maize, sunflower, durum wheat, moong etc., with around 0.1 million farmers under this program.

6.2.11 Promotion of Zero Tillage

The department of Agriculture, Government of Punjab is promoting "Zero Tillage Technique" since 2001–2002 in areas of state where wheat is sown after harvesting of rice. Zero till system refers to planting crops with minimum of soil disturbance. The other novel approach with much promise is the use of "Happy Seeder", which combines stubble mulching and seed drilling functions into one machine. The emphasis is on conserving moisture and residue management. Apart from benefits like proper mulching of paddy residue instead of burning, timely sowing, reducing run off and soil erosion, lesser deep percolation and improving soil health by incorporating plant nutrients, the zero tillage increases farmer's profit by Rs. 2,200–3,000/- per hectare by saving 80 % of diesel as wheat is sown in one pass only. The area under zero tillage in Punjab has increased from 6.83 thousand hectares in 2001–2002 to 412.69 thousand hectares in 2005–2006. As per the information provided by Punjab Agricultural University, zero tillage sowing of wheat on 412 thousand hectares in the state during rabi 2005–2006 has reduced the consumption of diesel by 20.6–24.7 million liters and also reduced the emission of CO_2 to the tune of 53.6–64.2 million kilograms in the environment on the basis of conversion factor of 2.6 kg of CO_2 per litre of diesel burnt.

6.2.12 Management of Agricultural Waste

Keeping in view the increasing problems associated with crop stubble burning, many initiatives have been taken to manage agricultural waste including paddy and wheat straw, cotton sticks, bagasse and animal waste. At present large volumes of wastes are being burnt in the field or as fuel. Several initiatives for its proper management have been taken up. These include:

6.2.13 Utilization of Straw and Husk

Though on paper, various district administrations in the state have imposed a ban on the burning of paddy straw in the fields after its harvesting, the problem still persists. As a result various departments and institutions are promoting alternative uses of straw instead of burning. These include:

6.2.14 Use of Rice Residue as Fodder for Animals

The rice residue as fodder for animals is not a very popular practice among farmers. This is mainly because of the high silica content in the rice residue. It is believed that almost 40 % of the wheat straw produced in the state is used as dry fodder for animals. However to encourage the use of rice residue as fodder for animals, a pilot project was taken up by PSCST at PAU under which trials on natural fermentation of paddy straw for use as protein enriched livestock feed were conducted. The cattle fed with this feed showed improvement in health and milk production. The technology was demonstrated in district Gurdaspur, Ludhiana, Hoshiarpur and Bathinda. The department of Animal Husbandry, Punjab has propagated the technology in the state.

6.2.15 Use of Crop Residue in Bio Thermal Power Plants

Another use of rice residue that is being encouraged by various institutions and departments is the use of rice residue for generation of electricity. The details of crop residue use in bio thermal plant is given in Chap. 4.

6.2.16 Use of Rice Residue as Bedding Material for Cattle

The farmers of the state have been advised to use paddy straw as bedding material for cross bred cows during winters as per results of a study conducted by the Department of Livestock Production and Management, College of Veterinary Sciences, Punjab Agricultural University. It has been found that the use of paddy straw bedding during winter helped in improving the quality and quantity of milk as it contributed to animals' comfort, under health and leg health. Paddy straw bedding helped the animals keep themselves warm and maintain reasonable rates of heat loss from the body. It also provides clean, hygienic, dry, comfortable and non-slippery environment, which prevents the chances of injury and lameness. Healthy legs and hooves ensure enhancement of milk production and reproductive efficiency of animals. The paddy straw used for bedding could be subsequently used in biogas plants. The use of paddy straw was also found to result in increased net profit of Rs. 188–Rs. 971 per animal per month from the sale of additional amount of milk produced by cows provided with bedding. The PAU has been demonstrating this technology to farmers through training courses, radio/TV talks and by distributing leaflets.

6.2.17 Use of Crop Residue for Mushroom Cultivation

Paddy straw can be used for the cultivation of *Agaricus bisporus*, *Volveriella volvacea* and *Pleurotus spp*. One kg of paddy straw yields 300, 120–150 and 600 g of these mushrooms respectively. At present, about 20,000 MT of straw is being used for cultivation of mushrooms in the state.

6.2.18 Use of Rice Residue in Paper Production

The paddy straw is also being used in conjunction with wheat straw in 40:60 ratio for paper production. The sludge can be subjected to biomethanization for energy production. The technology is already operational in some paper mills, which are meeting 60 % of their energy requirement through this method. Paddy straw is also used as an ideal raw material for paper and pulp board manufacturing. As per information provided by PAU, more than 50 % pulp board mills are using paddy straw as their raw material.

6.2.19 Use of Rice Residue for Making Bio Gas

The PSFC has been coordinating a project for processing of farm residue into biogas based on the technology developed by Sardar Patel Renewable Energy Research Institute (SPRERI). A power plant of 1 MW is proposed to be set up at Ladhowal on pilot basis on land provided by PAU. The new technology will generate 300 m^3 of biogas from one tonne of paddy straw.

6.2.20 Other Measures

For agricultural diversification, the new strategy lays emphasis on production of fruits and vegetables under controlled conditions, using modern practices like net houses, plastic tunnels and green houses. For achieving the same objective, half a million acres of land has been brought under crops, other than wheat and paddy, through contract farming. Yet another step towards diversification of agriculture, taken by the state Government, is the establishment of a new University of Animal Sciences. This is likely to impart desired impetus to dairy and livestock development. Besides, an Agriculture Diversification, Research and Development Fund, with an initial corpus of Rs. 20 crore have also been created. However, there is a need for creating a Venture Capital Fund at the National level for promoting agri-businesses.

Provision of Rs. 500 crore earmarked for the National Horticulture Mission during 2005–2006 is very meager, which needs to be enhanced to fully accommodate the requirement of funds for shifting from the traditional to horticulture crops.

While on the one hand there is an urgent need to revitalize the research in agriculture and related activities, on the other hand to tackle the problem of soil degradation and water depletion, a dedicated programme for promoting resource conservation technologies, such as zero tillage, deep ploughing, raised bed planting, laser land leveling etc., should be undertaken. Heavy investments are required to be made for rejuvenation of these resources. The Rashtriya Krishi Vikas Yojana (RKVY) is a welcome initiative.

The Central Pollution Control Board in exercise of its powers conferred under Section. 16(2) (h) of the Air (Prevention and Control of Pollution) Act, 1981(14 of 1981) hereby notify the National Ambient Air Quality Standards with immediate effect. Furthermore it is believed that the Punjab Government regularly publishes the adverse impacts of crop stubble burning in local newspapers. As per a local newspaper, the Tribune dated 19th May 2009, the District Magistrate Bhagwant Singh has banned the burning of crop stubble in Amritsar. However the practice still continues in the rural belt of Amritsar district, including Attari, Ajnala and Majitha.

The problem of pollution caused by rice and wheat crop stubble burning has not received much attention by the policymakers and the various pollution authorities. This could be partially due to the fact that the rice and wheat burning taken place only during selected months of October, November and December. The pollution is restricted only during these months. However even during these months there is considerable loss to human health and environment degradation. In the local dailies of Punjab you might come across articles requesting farmers to stop burning the stubble or creating awareness among them about its ill effects. But the problem still remains more or less unresolved. Though the Punjab Pollution Control Board has taken various measures to limit the amount of industrial pollution in the state of Punjab, however not much has been done to address agricultural pollution.

6.3 Summary of the Chapter

Punjab Government, its various Departments and other institutions like Punjab Agricultural University, Punjab Farmers Commission etc., are all making efforts to devise some alternate economic uses of rice stubble. These include the stubble treated with urea as a fodder for animals, its use in biothermal energy production, paper manufacturing, mushroom cultivation, bedding for animals, etc. Punjab government is also providing subsidy to the farmers to promote the use of equipments which help in checking the burning of crop residues, like rotavators, happy seeders, zero–till-drills and straw reapers. The Punjab Pollution Control Board (PPCB) has taken various measures to limit the amount of industrial pollution in the state but needs to do more to address agricultural pollution. Punjab State Council for Science and Technology is one of the institutes for imparting training on pollution control, waste management, clean technologies, environment policies, health monitoring and assessment and solid waste management conducted by the Central Pollution Control Board under the Human resource development programme. The Punjab Energy Development Agency was established in the year 1991, for the promotion and development of non-conventional and renewable energy programs or projects in the state of Punjab. Thus, as far as the institutional setup is concerned, there is enough constitutional provisions made under the law of the land to control and abet pollution related to agricultural waste burning. However, what is requisite to meet with this evil practice is strong will power among the governance and viable economic alternatives available to the farmers to keep the stubble burning practice at a bay.

Reference

Gupta, Prabhat, K., Sahai, Shivraj, Singh, Nahar, Dixit, C. K., Singh, D. P., Sharma, C., Tiwari, M. K., Gupta, Raj, K., & Garg, S. C. (2004). Residue burning in rice-wheat cropping system: causes and implications. *Current Science, 87*(12), 1713–1715 (Dec 25).

Chapter 7
Concluding Remarks and Policy Recommendations

Abstract Burning of farm waste causes severe pollution of land and water on local as well as regional scale. The off-field impacts are related to human health due to general air quality degradation resulting in aggravation of respiratory (like cough, asthma, bronchitis), eye and skin diseases. This study finds that total annual welfare loss in terms of health damages due to air pollution caused by the burning of paddy straw in rural Punjab amounts to Rs. 76 million. These estimates could be much higher if expenses on averting activities, productivity loss due to illness, monetary value of discomfort and utility could also be considered. To avoid burning of rice (and wheat) stubble, management of agricultural waste for alternate uses is being practiced and promoted. Various departments and institutions are promoting alternative uses of straw instead of burning, e.g., rice residue as fodder, crop residue in bio thermal power plants and mushroom cultivation, rice residue used as bedding material for cattle, production of bio-oil, paper production, bio-gas and in-situ. Other uses include incorporation of paddy straw in soil, energy technologies and its use in thermal combustion for generation of electricity.

Keywords Stubble burning · Alternate uses of rice stubble · Existing legislation on air pollution

7.1 Introduction

Agriculture sector is the prime mover of economic growth in Punjab. It has been governed by factors of production such as land, capital, energy, nutrients, water and other agricultural inputs. With only 1.5 % of geographical area of the country, Punjab has produced about 20 % of wheat, 10 % of rice and cotton each, of the aggregate produce of these crops in the country. The State is the chief granary of India contributing 22.1 % of rice and 38.7 % of wheat to the Central pool in 2011–2012. Further, over 95 % of the foodgrains moved inter-state to feed deficit areas through the Public Distribution System are the stocks procured from this State. It is characterized as the backbone of the Public Distribution System and a strong base for the food security of the country.

© The Author(s) 2015   133
P. Kumar et al., *Socioeconomic and Environmental Implications of Agricultural Residue Burning*, SpringerBriefs in Environmental Science,
DOI 10.1007/978-81-322-2014-5_7

Given tremendous achievements in the past, however serious concerns are now emerging about the future prospects of Punjab's agricultural sector. The greatest concern is about the over exploitation of ground water resources of the state. Further, in the past two to three decades, intensive agricultural practices have put a tremendous pressure on the soils and resulted in steady decline in its fertility (nutrient availability) both with respect to macro and micronutrients. Both rice and wheat have high nutritional requirements and the double cropping of this system has been heavily depleting the nutrient contents of soil. One of the recognized threats to the rice-wheat cropping system sustainability is the loss of soil organic matter as a result of rice-wheat residue burning in the fields.

Though various studies in the literature have addressed this issue of burning of the crop stubble but none have brought to the forefront the adverse implications of this unwarranted practice on human and animal health. The present study proceeds first by bringing to the forefront the amount of pollution being caused by rice residue burning. Thereafter the harmful effects of the pollution being generated by rice stubble burning on human and animal health are studied. Based on the information obtained, the study analyzes Punjab Government's existing policies to address the air pollution caused by rice stubble burning. Based on the findings of the Punjab Government policies to address the pollution caused by crop stubble burning, the study aims at providing policy suggestions to stamp out the practice.

7.2 Summary of the Findings

There is a misconception that industrial sector is the only contributor of pollution whereas agriculture also contributes to pollution in various ways. Crop residue burning is one among the many sources of air pollution. Due to technological advancements in the agricultural sector, waste concentration goes beyond certain limits thereby distorting the balance. Burning of farm waste causes severe pollution of land and water on local as well as regional scale. It is estimated that burning of paddy straw results in nutrient losses viz., 3.85 million tonnes of organic carbon, 59,000 t of nitrogen, 20,000 t of phosphorus and 34,000 t of potassium. This also adversely affects the nutrient budget in the soil. Straw carbon, nitrogen and sulphur are completely burnt and lost to the atmosphere in the process of burning. It results in the emission of smoke which if added to the gases present in the air like methane, nitrogen oxide and ammonia, can cause severe atmospheric pollution. These gaseous emissions can result in health risk, aggravating asthma, chronic bronchitis and decreased lung function. Burning of crop residue also contributes indirectly to the increased ozone pollution. It has adverse consequences on the quality of soil. When the crop residue is burnt the existing minerals present in the soil get destroyed which adversely hampers the cultivation of the next crop. Open field burning of crop stubble results in the emission of many harmful gases in the atmosphere, like Carbon Monoxide, N_2O, NO_2, SO_2, CH_4 along with particulate matter and hydro carbons.

However, appropriate assessments have not been undertaken to demonstrate the relevant impact of agriculturally based pollution to broad scale air pollution; that enforcement of the residue burning ban would lead to net public benefits; that adoption of the Happy Seeder is the most efficient means of utilizing/coping with the rice residue loads without resorting to burning; that the provision of government assistance is necessary or warranted to achieve efficient levels of Happy Seeder adoption; This study addresses these questions.

Crop residue burning is one among the many sources of air pollution. The on-field impact of burning includes removal of a large portion of the organic material, denying the soil an opportunity to enhance its organic matter and incorporate important chemicals such as nitrogen and phosphorus, as well as, loss of useful micro flora and fauna. The off-field impacts are related to human health due to general air quality degradation resulting in aggravation of respiratory (like cough, asthma, bronchitis), eye and skin diseases. Fine particles also can aggravate chronic heart and lung diseases and have been linked to premature deaths in people already suffering from these diseases. The black soot generated during burning also results in poor visibility which could lead to increased road side incidences of accident. It is thus essential to mitigate impacts due to the burning of agricultural waste in the open fields and its consequent effects on soil, ambient air and living organisms.

Several evaluations have been carried out to assess the impact of air pollution from agricultural residual burning. Air pollution contributes to the respiratory diseases like eye irritation, bronchitis, emphysema, asthma etc., which not only increase individuals' diseases mitigation expense but also affect their productivity at work. Most of the studies valuing health impacts of air pollution remain confined to urban areas as air pollution is considered mainly the problem of urban areas in developing countries. Though health consequences from burning of agricultural residue are not fully understood, relative short exposure may be more of a nuisance rather than a real health hazard. Many of the components of agricultural smoke cause health problem under certain conditions. There are many studies in developed countries that estimate the value of adverse health effects of air pollution. Similar evidences are available from India and other developing countries. These studies used either household health production model or damage function or cost of illness approaches to estimate the monetary value of health damage caused due to ambient air pollution. Note that these studies are restricted to measure the monetary value of reducing urban air pollution to the safe level since air pollution has been considered mainly the problem of urban areas.

These studies used either household health production model or damage function or cost of illness approaches to estimate the monetary value of health damage caused due to ambient air pollution. Note that these studies are restricted to measure the monetary value of reducing urban air pollution to the safe level since air pollution has been considered mainly the problem of urban areas. The present study used data of 625 individuals collected from a household level survey conducted in three villages, namely Dhanouri, Ajnoda Kalan and Simro of Patiala district of Punjab for 150 households. To get the estimates of monetary values of

human health impact of pollution two equations were estimated: one with mitigation expenditure and the other with workdays lost as dependent variables. Tobit
and Poisson models were used for estimating mitigation expenditure and workdays lost equations, respectively.

On an average, total amount of stubble generated for paddy and wheat per acre
was around 23 and 19 quintals, respectively. Out of this in the case of paddy, more
than 85 % was burnt in the open field and less than 10 % was incorporated, while
rest of 5 % was used for other purposes. In the case of wheat, 77 % of the total
amount was used as fodder for animals while 9 % was incorporated and around
11 % was burnt. Although farmers were convinced that burning was not harming
the level of crop yield but they pointed out that burning of field added extra cost to
the production because of top soil getting affected by the burning. The farmers who
burnt the field (fully or partly) to clear the wheat stubble used 169 kg of urea in the
next crop of paddy while those who incorporated or adopted other means used 145
and 148 kg of urea, respectively. Similarly, those farmers who burnt paddy field,
used added amount of Di-Amonia Phosphate (DAP) to recapture the nutritive lost
in the fire in comparison to those who incorporated or removed stubble manually.
Higher expenses were not only in terms of higher fertilizer but also in terms of
higher irrigation requirement by those who burnt their field to clear the stubble.

Household survey showed that paddy stubble burning leads to air pollution and
several other problems. Irritation in eyes and congestion in the chest were the two
major problems faced by the majority of the household members. Respiratory allergy,
asthma and bronchial problems were the other smoke related diseases which affected
household members in the selected villages. Almost 50 % of the selected households
indicated that their health related problems get aggravated during or shortly after
harvest when crop stubble burning is in full swing during the months of October,
November and December. In the peak season, affected families had to consult doctor or use some home medicine to get relief from irritation/itching in eyes, breathing
problem and similar other smoke related problems. On an average, the affected members suffered at least half a month from such problems and had to spend Rs. 300–
500 per household on medicine. In addition there were few examples where a family
member had to be hospitalized for three to four days and additional expenditure was
incurred. On an average, households spent around more than a thousand Rupees on
the non-chronic respiratory diseases like coughing, difficulty in breathing, irregular
heartbeat, itching in eyes decreased lung function etc., during the year 2008–2009.
However, out of this total expenditure, around 40–50 % was spent during the months
of October and November during the time of crop stubble burning. There was an additional cost in terms of household members remaining absent from work due to illness.

The study finds that total annual welfare loss in terms of health damages due to
air pollution caused by the burning of paddy straw in rural Punjab amounts to Rs. 76
million. These estimates could be much higher if expenses on averting activities, productivity loss due to illness, monetary value of discomfort and utility could also be
considered. There is additional monetary cost of burning to the farmers in terms of
additional fertilizer, pesticides and irrigation. One also needs to add the losses of soil
nutrient, vegetation, bio-diversity and accidents caused because of low visibility.

To avoid burning of rice (and wheat) stubble, management of agricultural waste for alternate uses is being practiced and promoted. Agricultural waste includes paddy and wheat straw, cotton sticks, bagasse and animal waste. Keeping in view the increasing problems associated with crop stubble burning several initiatives for its proper management have been taken up. Various departments and institutions are promoting alternative uses of straw instead of burning. These include use of rice residue as fodder, crop residue in Bio thermal power plants and mushroom cultivation, rice residue used as bedding material for cattle, production of bio-oil, paper production, bio-gas and in-situ. Other uses include incorporation of paddy straw in soil, energy technologies and thermal combustion.

The problem of pollution caused by rice and wheat crop stubble burning has not received much attention by the policymakers and the various pollution authorities till recently. This could be partially due to the fact that the rice burning (the major source agri waste burning pollution) takes place only during selected months of October, November and December. The pollution is restricted only during these months. However even during these months there is considerable loss to human health and environment degradation. It is believed that Punjab Government regularly publishes the adverse impacts of crop stubble burning in local newspapers. Punjab Government, its various Departments and other institutions like Punjab Agricultural University, Punjab Farmers Commission etc., are all making efforts to devise some alternate economic uses of rice stubble. These include the stubble treated with urea as a fodder for animals, its use in biothermal energy production, paper manufacturing, mushroom cultivation, bedding for animals, etc. Punjab government is also providing subsidy to the farmers to promote the use of equipments which help in checking the burning of crop residues, like rotavators, happy seeders, zero–till-drills and straw reapers. In the local dailies of Punjab one comes across articles requesting farmers to stop burning the stubble or creating awareness among them about its ill effects. An example, the District Magistrate Amritsar banned the burning of crop stubble (the Tribune dated 19th May 2009). However the practice still continues in the rural belt of Amritsar district, including Attari, Ajnala and Majitha. Thus the problem of agri waste burning still remains unresolved. While on the one hand, there is an urgent need to revitalize the research in agriculture and related activities, on the other hand, to tackle the problem of soil degradation and water depletion, a dedicated programme for promoting resource conservation technologies, such as zero tillage, deep ploughing, raised bed planting, laser land leveling etc., should be undertaken. Heavy investments are required to be made for rejuvenation of these resources. The Rashtriya Krishi Vikas Yojana (RKVY) is a welcome initiative in that direction. There is a requirement for an eco-friendly technology that will be beneficial to the farmer community and the State by providing them a tool for improving soil health and environment for sustainable agriculture.

India is a legislation rich country with reference to pollution. The Ministry of Forest and Environment is a vital agency in the administrative setup of the Union Government. The Ministry is entrusted with the task of planning, coordinating, overseeing and implementing various forestry and environment programmes. In order to promote the development of clean technology, development of tools and techniques for pollution prevention and to formulate sustainable development strategies, the Ministry granted an aid in 1994 for the development and promotion of clean technologies. Eleven major laws exist to control pollution in India and many forums for their implementation in various ways. Among the existing legislation on air pollution in India includes: Air Prevention and Control of Pollution Act, 1981; The Environment Protection Act, 1986; The National Environment Tribunal Act, 1995; The National Environment Appellate Authority Act, 1997; and Biological Diversity Act, 2002. Under these different Acts, provisions are made to protect the environment from all kinds of pollution related to industrial and agricultural activities. The Punjab Pollution Control Board (PPCB) is entrusted with the functions of planning a comprehensive program for the prevention, control and abatement of pollution in Punjab. PPCB has to support and encourage developments in the field of pollution control. PPCB has taken various measures to limit the amount of industrial pollution in the state but not much has been done to address agricultural pollution. The Central Pollution Control Board (CPCB) is the 'Central Board' for the prevention, control and abatement of air and water pollution in India. CPCB has initiated a nationwide programme of ambient air quality monitoring called NAMP.

The division of environment assists the State Department of Environment, Government of Punjab in technical matters pertaining to environment, identification of major areas of ecological concern, defining the State Government policies and plans on various environmental issues, coordinating and monitoring schemes related to environment, creating environmental awareness and promoting environmental education, training and research.

Moreover Punjab State Council for Science and Technology was also recognized as one of the institutes for imparting training on pollution control, waste management, clean technologies, environment policies, health monitoring and assessment and solid waste management conducted by the Central Pollution Control Board under the Human resource development programme. The Punjab Energy Development Agency was established in the year 1991, for the promotion and development of non-conventional and renewable energy programs or projects in the state of Punjab. Thus, as far as the institutional setup is concerned, there is enough constitutional provisions made under the law of the land to control and abet pollution related to agricultural waste burning. However, what is requisite to meet with this evil practice is strong will power among the governance and viable economic alternatives available to the farmers to keep the stubble burning practice at a bay.

7.3 Policy Recommendations and Research Needs

- The burning problem is rampant in Punjab, Haryana and Uttar Pradesh and data about air pollution in rural areas is too scarce as most of the pollution monitoring stations are setup in urban areas. Therefore, it is necessary to have a proper idea of real amount of air pollution generated by the burning of crop residuals. It is also necessary to have exact measurement of RSPM, black carbon (soot) in the ambient air, measurement of meteorological parameters like wind velocity, temperature profile and humidity etc., in the rural areas to initiate policy actions to avoid the same.

- Imposing ban on burning legally may not succeed unless farmers are properly educated and made aware about its adverse implications for human and animal health and its undesirable impact on soil, biodiversity etc. To educate farmers, extension activities like Documentary on environment and climate change may be made. In the documentary emphasis should be put on how burning adversely impact the climate change and educate the farmers about the economics of not burning the agricultural residues.

- Alternatives to burning agricultural residue like collection and transportation of agricultural residues, gasification as a fuel for the boilers, converting into briquettes and designing of suitable harvester should be promoted.

- Free electricity should not be promoted as the same policy has led to installation of high powered tube wells that are responsible for over draw water from deep inside the earth.

- In-situ management in the field, composting by chemical means and straw mulching by mechanical means should be promoted. The machines like the use of disc plough, disc harrow, rotavator, zero tillage and happy seeder can help in mulching the crop stubble.

- Wastes/residue should be collected from the fields and should be disposed off or used for making useful products like making compost, organic manure to improve soil fertility, and gasification for use as a fuel or for power generation; night soil to produce biogas and manure.

- The stem may be cut from the root level itself. The same would require a suitable thresher cum harvester that should be developed using indigenous techniques. Use high power tractor for deep cutting. For small farmers it can be followed on cooperative basis.

- Make the small farmers to understand that making chaff out of the agricultural residues is to their advantage.

Annexure

Schedule 1

Following is the list of industries requiring environment clearance from the Central Government.

- Nuclear Power and related projects such as Heavy Water Plants, nuclear fuel complex, rare earths.
- River Valley projects including hydel power projects, major irrigation projects and their combination including flood control project except project relating to improvement work including widening and strengthening of existing canals with land acquisition up to maximum of 20 m (on both sides put together) along the existing alignment provided such canals do not pass through ecological sensitive areas such as natural parks, sanctuaries, tiger reserves and reserve forests.
- Ports, Harbors, Airports (except minor ports and harbors).
- Petroleum Refineries including crude and product pipelines, isolated petroleum product storages.
- Chemical Fertilizers (Nitrogenous and Phosphatic other than single super phosphate).
- Pesticides (Technical).
- Petrochemical complexes (Both Olefinc and Aromatic) and Petro-chemical intermediate such as DMT. Caprolactam, LAB etc. and production of basic plastics such as LDPE, HDPE, PP, PVC.
- Bulk drugs and pharmaceuticals.
- Exploration for oil and gas and their production, transportation and storage.
- Synthetic Rubber.
- Asbestos and Asbestos products.
- Hydrocyanic acid and its derivatives.
- Primary metallurgical industries (such as production of Iron and Steel, Aluminum, Copper, Zinc, Lead and Ferro Alloys) (a) Electric are furnaces (Mini Steel Plants).
- Chlor alkali industry.

© The Author(s) 2015
P. Kumar et al., *Socioeconomic and Environmental Implications of Agricultural Residue Burning*, SpringerBriefs in Environmental Science,
DOI 10.1007/978-81-322-2014-5

- Integrated paint complex including manufacture of resins and basic raw materials required in the manufacture of paints.
- Viscose Staple fiber and filament yarn.
- Storage batteries integrated with manufacture of oxides of lead and lead an antimony alloy.
- All tourism projects between 200–500 m of High Tide Line or at locations with an elevation of more than 1000 m with investment of more than Rs. 5 crores.
- Thermal Power Plants.
- Mining projects (major minerals) with leases more than 5 ha.
- Highway Projects except projects relating to improvement work including widening and strengthening of roads with marginal land acquisition along the existing alignment provided it does not pass through ecological sensitive areas such as National Parks, Sanctuaries, Tiger Reserves, reserve forests.
- Tarred Roads in Himalayas and/or Forest areas.
- Distilleries.
- Raw Skins and Hides.
- Pulp, paper and newsprint.
- Dyes.
- Cement.
- Foundries (individual).
- Electroplating.
- Meta Amino Phenol.
- New Construction projects.
- New Industrial Estates.

Schedule 2

The industries under the **Red Category** are:

- Distillery including Fermentation industry.
- Sugar (excluding Khandsari).
- Fertilizer.
- Pulp and Paper (Paper manufacturing with or without pulping).
- Chlor alkali.
- Pharmaceuticals (Basic) (excluding formulation).
- Dyes and Dye-intermediates.
- Pesticides (Technical) (excluding formulation).
- Oil refinery (Mineral oil or Petro refineries).
- Tanneries.
- Petrochemicals (Manufacture of and not merely use of as raw material).
- Cement.
- Thermal power plants.
- Iron and Steel (Involving processing from ore/scrap/Integrated steel plants).

- Zinc smelter.
- Copper smelter.
- Aluminum smelter.
- Tyres and tubes (Vulcanisation/Retreading/moulding).
- Synthetic rubber.
- Glass and fiber glass production and processing.
- Industrial carbon including electrodes and graphite blocks, activated carbon, carbon black etc.
- Paints and varnishes (excluding blending/mixing).
- Pigments and intermediates.
- Synthetic resins.
- Petroleum products involving storage, transfer or processing.
- Lubricating oils, greases or petroleum—based products.
- Synthetic fiber including rayon, tyre cord, polyester filament yarn.
- Surgical and medical products involving prophylactics and latex.
- Synthetic detergent and soap.
- Photographic films and chemicals.
- Chemical, petrochemical and electrochemical including manufacture of acids such as Sulphuric Acid, Nitric Acid, Phosphoric Acid etc.
- Industrial or inorganic gases.
- Chlorates, per chlorates and peroxides.
- Glue and gelatin.
- Yarn and textile processing involving scouring, bleaching, dyeing, printing or any effluent/emission generating process.
- Vegetable oils including solvent extracted oils, hydro-generated oils.
- Industry or process involving metal treatment or process such as picking, surface coating, paint baking, paint stripping, heat treatment, phosphate or finishing etc.
- Industry or process involving electroplating operations.
- Asbestos and asbestos-based industries.
- Slaughter houses and meat processing units.
- Fermentation industry including manufacture of yeast, beer etc.
- Steel and steel products including coke plants involving use of any of the equipment's such as blast furnaces, open hearth furnace, induction furnace.
- Incineration plants.
- Power generating plants (excluding D.G. Sets).
- Lime manufacturing.
- Tobacco products including cigarettes and tobacco processing.
- Dry coat processing/Mineral processing industries like ore sintering, etc.
- Phosphate rock processing plants.
- Coke making, coal liquefaction, coal tar distillation or fuel gas making.
- Phosphorous and its compounds.
- Explosives including detonators, fuses etc.
- Fire crackers.
- Processes involving chlorinated hydrocarbons.
- Chlorine, fluorine, bromine, iodine and their compounds.

- Hydrocyanic acid and its derivatives.
- Milk processing and dairy products (Integrated Project).
- Industry or process involving foundry operations.
- Potable alcohol (IMFL) by blending or distillation of alcohol.
- Anodizing.
- Ceramic/refractories.
- Lead processing and battery reconditioning and manufacturing including lead smelting.
- Hot Mix plants.
- Hospitals.
- Mining and ore-beneficiation.

The following industries fall under the '**Orange Category**'

- Manufacture of mirror from sheet glass and photo framing.
- Cotton spinning and weaving.
- Automobile servicing and repairs stations.
- Hotels and restaurants.
- Flour mills (excluding Domestic Atta Chakki).
- Malted food.
- Food including fruits and vegetable processing.
- Pulping and fermenting of coffee beans.
- Instant tea/coffee, coffee processing.
- Non-alcoholic beverages (soft drinks).
- Fragrances and industrial perfumes.
- Food additives, nutrients and flavors.
- Fish processing.
- Organic nutrients.
- Surgical and medical products not involving effluent/emission generating processes.
- Laboratory-wares.
- Wire drawing (cold process) and bailing straps.
- Stone crushers.
- Laboratory chemicals involving distillation, purification process.
- Tyres and tubes vulcanization, molding.
- Pesticides/Insecticides/Fungicides/Herbicides/Agro chemical formulation.
- NPK Fertilizers/Granulation.
- Pharmaceuticals formulation.
- Khandsari sugar.
- Pulverizing units.